S0-GQF-621

UNIVERSITY OF MAINE

RAYMOND H. FOGLER LIBRARY

Improvement of Cereal Quality by Genetic Engineering

Improvement of Cereal Quality by Genetic Engineering

Edited by

Robert J. Henry
Queensland Agricultural Biotechnology Centre
University of Queensland
St. Lucia, Queensland, Australia

and

John A. Ronalds
Division of Plant Industry
CSIRO
North Ryde, NSW, Australia

Plenum Press • New York and London

Library of Congress Cataloging in Publication Data

Improvement of cereal quality by genetic engineering / edited by Robert J. Henry and John A. Ronalds.

p. cm.

"Proceedings of the Royal Australian Chemical Institute, Cereal Chemistry Division Symposium on Improvement of Cereal Quality by Genetic Engineering, held September 12–16, 1993 . . . at Sydney, Australia"—T.p. verso.

Includes bibliographical references and index.

ISBN 0-306-44721-5

1. Grain—Genetic engineering—Congresses. 2. Grain—Quality—Congresses. I. Henry, Robert J. II. Ronalds, John A. III. Royal Australian Chemical Institute. Cereal Chemistry Division. IV. Symposium on Improvement of Cereal Quality by Genetic Engineering (1993: Sydney, N.S.W.)

SB189.53.I57 1994 94-8261

633.1'0423–dc20 CIP

Proceedings of the Royal Australian Chemical Institute, Cereal Chemistry Division Symposium on Improvement of Cereal Quality by Genetic Engineering, held September 12–16, 1993, during the Guthrie Centenary Conference at Coogee Beach, Sydney, Australia

ISBN 0-306-44721-5

©1994 Plenum Press, New York
A Division of Plenum Publishing Corporation
233 Spring Street, New York, N.Y. 10013

All rights reserved

No part of this book may be reproduced, stored in a retrieval system, or transmitted in any form or by any means, electronic, mechanical, photocopying, microfilming, recording, or otherwise, without written permission from the Publisher

Printed in the United States of America

PREFACE

If I had to nominate an area of food production in which science has played a major role in addressing product quality to meet market needs I would not pass by the intimate relationship of cereal chemistry with cereal plant breeding programs.

In Australia, cereal chemistry and product quality labs have long been associated with wheat and barley breeding programs. Grain quality characteristics have been principal factors determining registration of new cultivars.

This has not been without pain in Australia. On the one hand some cultivars with promising yield and agronomic characteristics have been rejected on the basis of quality characteristics, and for a period our breeders imposed selection regimes based on yield which resulted in declining quality characteristics.

In the end the market provides the critical signals.

For many years Australia held a commanding market position on the basis of a single quality image, initially based on bulked wheat of fair/average quality (FAQ). Later this was improved by segregation into four broad classes* based around Australian Standard White (ASW). This is no longer a viable marketing strategy. We were probably a little slow in recognising the mosaic of present day wheat markets, but now have up to 18 different grades available.

Around the world wheat is a grain with many end uses. Its use in bread is expanding. I could not help noticing the number of hot bread shops in Japan when I was there last week and at a dinner I was intrigued that all the Western visitors ordered rice and all our Japanese hosts ordered bread!

There are also expanding markets, as a result of the internationalisation and interdigitation of the various cultures in the world, in the use of wheat in different types of noodles and flat breads.

Australia, like any other wheat producing country, must decide which market, or markets, it is aiming at, and produce the grain that is ideally suited to those markets. This is where cereal chemistry research into quality characteristics is critical.

Also with barley, Australia has made mistakes. In recent years the preoccupation of our breeders with yield has resulted in lack of improvement in malting quality characteristics and we lost markets to other producer countries where breeders had read the market signals more astutely. This difficulty was exacerbated in Australia by failure to recognise the need to segregate barley. For too long we have tried to produce an FAQ barley, suited to both the feed and the malting markets.

---

* 1. Australian Prime Hard. 2. Australian Hard. 3. Australian Standard White. 4. Australian Soft.

Just as we were slow to recognise some of the market signals in our barley and wheat improvement programs, so too have we been conservative in our approach to analysis and adjustment of quality characteristics.

We have strength in cereal chemistry in this country and have made contributions which have been recognised and used around the world, but our cereal chemistry, in my view, has not kept pace with developments in cellular and molecular biochemistry, which is what cereal chemistry is really about. We need to recognise that we must add new tools and new experimental approaches to our cereal quality research.

The baking industry and its technocracy has also been unduly conservative. The machines with which we measure critical parameters of dough are not the machines one would expect with the level of sophistication now possible in instrumentation.

Too long our chemists have grappled with the extraordinarily difficult problems of dealing en masse with the major proteins which are components in flour. We have not had adequate methods for determining the properties of individual proteins and then set about with a rational plan to make specific changes to the complement of proteins, in order to provide an improvement in a character like dough strength.

Now we have new technologies which will radically improve our capacities for analysis and adjustment of product quality. You have recognised this in putting together this special symposium on the improvement of cereal quality by genetic engineering.

Genetic engineering of cereals is now a real prospect. These recalcitrant monocots have now succumbed to science and gene transfer systems are in place for all the major cereals.

Rice transformation was reported in 1988 and is now carried out in many laboratories, maize followed and although still difficult transformation no longer presents a barrier for maize improvement.

This symposium is one of the first in which convincing reports of wheat and barley transformation will be given. Bioloistics has been an important technological development to help bring this about. From now on genetic engineering should be an integral technology in cereal breeding programs.

Genetic engineering is really the ultimate form of limited back cross breeding and will have a role in precise adjustment and addition of traits to top cultivars. In rice the early emphasis has been for genetic engineering to be focused on insect pests and viral diseases, not so much on quality characteristics of the grain. But in both barley and wheat quality traits are likely to be the foci of much of the earliest genetic engineering research. This is because molecular biology is presenting us with ways of tackling well defined problems in these two cereals. We can expect to see seed proteins of wheat come under intense biochemical and biophysical analysis and we will produce wheats with precisely adjusted protein portfolios. The beautiful work of Jack Preiss, collaborating with Monsanto scientists on altering the biosynthesis of starch in potato, also opens for us a door too long closed and ignored in wheat.

In malting barley one of the major outcomes of genetic engineering will be that the maltsters themselves will be better able to define exactly what they want improved!

In genetic engineering a key need is the control of expression of any transgene. As much attention has to be given to the promoter controls as to the coding region in any gene construct. Time, place and quantity of expression depends upon the interactions of a number of transcription factors, activators and enhancers. Our knowledge of these upstream sequences and their binding proteins is increasing rapidly and so, too, is the quality of performance of transgenes. Exciting new tools are being developed, for example plantibodies may give us new means of crop protection but may also provide us with exquisitely powerful tools for intracellular analysis.

What else is in store for us?

I have emphasised the precision of adjustment of quality traits. Breeders will have a much increased pool of variation, and there will be novel characters giving us new opportunities with cereals. Cereals may, in the future, be valuable sources of products for the pharmaceutical industry and for the enzymes and feedstocks of future industries. I am sure biological capability will be there. Again, market forces will ultimately decide. But along with the market will be another significant determining force, that of societal acceptance of this new technology. I am pleased to see that this is a topic that you are treating, along with scientific developments, in this exciting conference.

J. Peacock
CSIRO
Division of Plant Industry
13 September 1993

ACKNOWLEDGMENTS

This symposium on the improvement of cereal quality by genetic engineering was initiated by the Cereal Chemistry Division of the Royal Australian Chemical Institute. Funds to support the symposium came mainly from the Cereal Chemistry Division and the Grains Research and Development Corporation. The symposium was held as part of the Guthrie Centenary Conference which was the 43rd annual cereal chemistry conference held by the Cereal Chemistry Division in Australia. The Guthrie Centenary Conference celebrated the centenary of cereal chemistry in Australia, it being 100 years since the pioneering chemist Frederick Guthrie began his historic collaboration with the wheat breeder, William Farrer. The symposium demonstrated that genetic engineering is now poised to take the discipline of cereal chemistry into a new era. The symposium was made possible by the efforts of the conference organising committee; J Ronalds, W Sing, J Camilleri, R Henry, B Rose, L Welsh and M Wootton and by the contributors of research papers.

CONTENTS

SECTION I
DEVELOPMENT OF TECHNIQUES FOR TRANSFORMATION OF CEREALS

SECTION II
GENETIC ENGINEERING OF CEREAL PROTEIN QUALITY

SECTION III
GENETIC ENGINEERING OF CEREAL STARCH QUALITY

SECTION IV
IMPROVEMENT OF BARLEY QUALITY BY GENETIC ENGINEERING

SECTION V
REGULATION OF CEREAL GENETIC ENGINEERING

SECTION I

DEVELOPMENT OF TECHNIQUES FOR TRANSFORMATION OF CEREALS

ASSESSMENT OF METHODS FOR THE GENETIC TRANSFORMATION OF WHEAT

R. I. S. Brettell, D. A. Chamberlain, A. M. Drew, D. McElroy, B. Witrzens and E. S. Dennis

CSIRO Division of Plant Industry
GPO Box 1600
Canberra City, ACT 2601
Australia

SUMMARY

In the search for a routine and reliable method for wheat transformation, a wide range of approaches have been tried. Some of these have not withstood rigorous testing, and it appears that many early reports of wheat transformation may have resulted from experimental artefact. We have evaluated two methods for wheat transformation: direct gene transfer to protoplasts and transformation of intact tissues with DNA-coated microparticles. While both methods have yielded transformed wheat tissue, each has its advantages and disadvantages. On balance, microparticle bombardment of freshly initiated cultures is the preferred method for wheat transformation as it avoids the necessity of establishing the longer term embryogenic cultures required for protoplast isolation, and it results in the recovery of a high proportion of phenotypically normal plants from the transformed materials.

INTRODUCTION

Transformation of wheat is a prerequisite for genetic engineering of this important crop species. The aim of this paper is to examine the range of techniques that have been used in attempts to achieve genetic transformation of cereals, and thereby provide an indication of the most suitable method for the reliable production of transformed wheat plants.

Ten years have elapsed since the first reports of transformation appeared for broad-leaved plant species, yet progress with grasses and cereals has been relatively slow (Potrykus, 1990). This can be attributed to two features of graminaceous plants which have hindered the application of some of the more commonly used routes to plant transformation. First is a lack of response to gene transfer techniques mediated by *Agrobacterium tumefa-*

Improvement of Cereal Quality by Genetic Engineering, Edited by
Robert J. Henry and John A. Ronalds, Plenum Press, New York, 1994

ciens. While it has been demonstrated that infectious viral sequences can be introduced into monocots through a process known as agroinfection (Grimsley et al., 1987; Dale et al., 1989), there are only isolated reports of stable transformation of cereal cells through the integration of T-DNA from the Ti-plasmid of *Agrobacterium* (Deng et al., 1990; Raineri et al., 1990; Gould et al., 1991; Mooney et al., 1991). These experiments have not been satisfactorily reproduced and so far do not provide sufficient encouragement that *Agrobacterium* can be used as a reliable vector for cereal transformation.

The second barrier to cereal transformation is associated with difficulties in recovering plants from isolated protoplasts of graminaceous monocots. Efficient methods have been developed for direct gene transfer into monocot protoplasts (Potrykus et al., 1985; Davey et al., 1989), however recovery of transformed plants depends on efficient protoplast regeneration which until recently has been lacking for most cereal species.

Plant cells present many obstacles to the introduction of exogenous DNA, not least of which is the presence of a thick cellulosic wall which is impervious to larger molecules such as nucleic acids. In plants, extracellular nucleases are often produced in abundance and these act to degrade any unprotected DNA molecules. There are also structural constraints imposed by small nuclear and cytoplasmic volumes, in many cell types, associated with the presence of large vacuoles and other microbodies. In the absence of a transformation system based on *Agrobacterium*, a number of approaches have been tried in order to overcome these barriers and provide a system wherby a proportion of target cells will have DNA molecules reaching the nucleus where integration can occur. Among the methods applied to cereal transformation are injection of developing tillers with DNA solutions (De la Pena et al., 1987), the 'pollen-tube pathway' for which DNA is applied to florets near the time at which pollination occurs (Luo and Wu, 1988), treating floral organs with *Agrobacterium* (Hess et al., 1990), and electrophoresis of DNA into seed tissues (Ahokas, 1989). For the most part, these methods have been spectacularly unsuccessful. The experiments have lacked reproducibility, and the evidence for integration of DNA into the plant genome has been poor. Some of the results may have been due to artefact, such as transformation of endophytic microorganisms as suggested by Langridge et al. (1992). One notable exception to these newer methods has been bombardment of cells with DNA-coated microparticles. This method was first used successfully to transform tobacco cells (Klein et al., 1988) and has since been extended to transform plants in many different genera.

The relatively low frequencies of DNA delivery achieved in most cases mean that methods have had to be devised to allow efficient selection of cells which carry and express introduced gene sequences. The selection regimes are generally based on the expression of a marker gene producing an enzyme which confers resistance to a cytotoxic substance, usually an antibiotic or herbicide. In this paper the results we have obtained for wheat using microparticle bombardment will be compared to those from experiments using direct gene transfer to protoplasts.

DIRECT GENE TRANSFER TO PROTOPLASTS

Protoplasts are single cells from which the cell wall has been removed. This is generally achieved by digesting plant tissues in a solution containing a mixture of cellulytic and pectolytic enzymes. Plants were first regenerated from tobacco protoplasts more than 20 years ago (Nagata and Takebe, 1971), however it is only in recent times that the technique has been extended to include cereal species, notably rice (Abdullah et al., 1986; Yamada et al., 1986; Kyozuka et al., 1987).

DNA can be introduced into isolated protoplasts following treatment with poyethylene glycol (Krens et al., 1982; Potrykus et al., 1985; Maas and Werr, 1989). Alternatively the protoplasts can be subjected to electroporation where an electric field facilitates uptake of DNA through the protoplast plasma membrane (Fromm et al., 1985; Nagata, 1989; Larkin et al., 1990). Transformation of plants by direct gene transfer to protoplasts was first achieved with tobacco (Paszkowski et al., 1984). Following the development of methods to regenerate plants from cultured cereal protoplasts, this approach resulted in the production of transgenic rice (Toriyama et al., 1988; Shimamoto et al., 1989; Zhang et al., 1988) and maize (Rhodes et al., 1988).

Protoplasts can provide a large and uniform population of target cells for the introduction of DNA, however application of the method to cereal transformation is limited by difficulties encountered in obtaining sustained divisions and plant regeneration from cultured protoplasts. In cereals, regeneration of plants has only been achieved from protoplasts isolated from embryogenic suspension cultures. There are no confirmed reports of regeneration of plants from protoplasts isolated from an intact cereal plant. Embryogenic suspension cultures are difficult to establsh and maintain. Although successful regeneration of wheat plants from protoplasts has now been achieved independently in a number of laboratories (Harris et al., 1988; Ren et al., 1989; Vasil et al., 1990; Chang et al, 1991; He et al., 1992; Li et al., 1992; Qiao et al., 1992), mature, fertile plants have been produced infrequently. In part this may be due the length of time required to establish embryogenic cultures suitable for cereal protoplast isolation. The longer the time wheat cells are in culture, the greater is the likelihood that deleterious mutations will be present in the regenerating plant materials.

Initiation of embryogenic suspension cultures in wheat appears to be considerably more difficult than for rice. While we have been able to initiate embryogenic suspensions for a small number of hexaploid wheat genotypes (derived from the cultivars Sunstar, Millewa and Vilmorin 27) and select for transformed cultures following electroporation of protoplasts, only weak and chlorophyll deficient shoots have beeen regenerated from these cultures (D.A. Chamberlain, A. Drew and B. Witrzens, unpublished data). This is in contrast to our experience with rice for which we have been able to obtain fertile transgenic plants for three genotypes (Oryza sativa cvs. Taipei 309, Kinuhikari and Calrose) by direct gene transfer to protoplasts (B. Witrzens and R.I.S. Brettell, unpublished data; Chamberlain et al., 1994).

Further progress is being made in regenerating plants from transformed protoplasts of bread wheat (see K.J Scott, D.G. He, S. Karunaratne and A. Muradov, this volume) and its progenitor species such as *Triticum tauschii* (S.Sterle and J. Kollmorgen, personal communication). However, the efficiency of the process may need to be improved before direct gene transfer to protoplasts can provide a routine method for wheat transformation.

MICROPARTICLE BOMBARDMENT OF REGENERABLE TISSUES

The technique of microparticle bombardment or biolistics is a relatively recent innovation, yet it has shown considerable versatility and has been used to transform a wide range of plant species, including cereals such as maize (Gordon-Kamm et al., 1990; Fromm et al., 1990), rice (Christou et al., 1991), wheat (Vasil et al., 1992) and oats (Somers et al., 1992). Introduction of DNA into target cells is mediated by small particles of tungsten or gold which are sufficiently small to penetrate individual cells without destroying their integrity and viability (Sanford et al., 1987)

Microparticle bombardment has an advantage over direct gene transfer to protoplasts that DNA can be directly introduced into cells in intact tissues from which plants can readily be regenerated. However, potentially transformed cells will be connected to and surrounded by non-transformed cells in the multicellular target tissue. Therefore the successful application of the method depends on an efficient and discriminating means of selecting for the proliferation of those cells which carry the exogenous DNA. As with direct gene transfer to protoplasts selection has been achieved using antibiotics such as kanamycin, geneticin and hygromycin (Vasil et al., 1991; Bower and Birch, 1992; Li et al., 1993) following delivery of DNA carrying the corresponding gene for resistance. However, increasingly, selections are being made with genes for herbicide resistance. These include the *bar* gene from *Streptomyces hygroscopicus* which codes for phosphinothricin N-acetyl transferase (PAT), conferring resistance to phosphinothricin and bialaphos, and mutant plant acetolactate synthase (*als*) genes which confer resistance to sulphonylurea herbicides such as chlorsulfuron and sulfometuron methyl (Wilmink and Dons, 1993). For example, we have recovered transformed wheat callus following bombardment of suspension cultures with tungsten particles coated with plasmid DNA encoding a mutant *als* gene from tobacco and selection with sulfometuron methyl (Chamberlain et al., 1994). More recently, we have used the technique to recover fertile wheat plants following microparticle bombardment of cultured immature embryos using the same herbicide selection (R.I.S. Brettell, D.A. Chamberlain and D. McElroy, unpublished results).

The effectiveness of this approach has been demonstrated by the other contributors to this volume, who have shown the application of microparticle bombardment for transformation of wheat, barley and oats. A common thread from these data is that transformed plants can be obtained within a relatively short period. Moreover, by avoiding long term embryogenic cultures of the type that were initially used in wheat transformation (Vasil et al., 1991, 1992), the majority of plants obtained are phenotypically normal and retain a high degree of fertility.

CONCLUDING REMARKS

Cereal transformation is a fast evolving field, and new methods are being developed which may offer promise for the future. One example involves DNA delivery by electroporation of intact tissues (D'Halluin et al., 1992) which has been applied to generate transformed maize plants and is currently being evaluated for other cereal species. As with other methods, the usefulness of this approach cannot be assessed until the method has been satisfactorily reproduced in other laboratories.

With the information currently to hand, it appears that microparticle bombardment or biolistics offers the best prospect for providing a reliable method for wheat transformation. Emphasis can now be placed on further optimisation of all aspects of the procedure, including DNA delivery, selection of transformed tissue and regeneration of plants.

A routine method for wheat transformation finally provides a means of testing the behaviour of introduced gene sequences in wheat. In particular the stability of gene expression needs to be monitored before practical application of the technology can be considered.

Acknowledgements

This project was supported by the Wheat Committee of the Grains Research and Development Corporation of Australia.

REFERENCES

Abdullah, R., Cocking, E.C., and Thompson, J.A. (1986) Efficient plant regeneration from rice protoplasts through somatic embryogenesis. *Bio/Technology* 4, 1087–1090.

Ahokas, H. (1989) Transfection of germinating barley seed electrophoretically with exogenous DNA. *Theor. Appl. Genet.* 77, 469–472.

Bower, R. and Birch, R.G. (1992) Transgenic sugarcane plants via microprojectile bombardment. *The Plant Journal* 2, 409–416.

Chamberlain, D.A., Brettell, R.I.S., Last, D.I., Witrzens, B., McElroy, D., and Dennis, E.S. (1994). The use of the Emu promoter with antibiotic and herbicide resistance genes for the selection of transgenic wheat callus and rice plants. *Aust. J. Plant Physiol.* 21, 95–112.

Chang, Y.F., Wang, W.C., Warfield, C.Y., Nguyen, H.T., and Wong, J.R. (1991) Plant regeneration from protoplasts isolated from long-term cell cultures of wheat (*Triticum aestivum* L.). *Plant Cell Reports* 9, 611–614.

Christou, P., Ford, T.L., and Kofron, M. (1991) Production of transgenic rice (*Oryza sativa* L.) plants from agronomically important indica and japonica varieties via electric discharge particle acceleration of exogenous DNA into immature zygotic embryos. *Bio/Technology* 9, 957–962.

Dale, P.J., Marks, M.S., Brown, M.M., Woolston, C.J., Gunn, H.V., Mullineaux, P.M., Lewis, D.M., Kemp, J.M., Chen, D.F., Gilmour, D.M., and Flavell, R.B. (1989) Agroinfectionof wheat: inoculation of in vitro grown seedlings and embryos. *Plant Science* 63, 237–245.

Davey, M.R., Rech, E.L., and Mulligan B.J. (1989) Direct DNA transfer to plant cells. *Plant Mol. Biol.* 13, 273–285.

De la Pena, A., Loerz, H., and Schell, J. (1987) Transgenic plants obtained by injecting DNA into young floral tillers. *Nature* 325, 274–276.

Deng, W.Y., Lin, X.Y., and Shao, Q.Q. (1990) *Agrobacterium tumefaciens* can transform *Triticum aestivum* and *Hordeum vulgare* of Gramineae. *Science in China* (Series B) 33, 27–34.

D'Halluin, K., Bonne, E., Bossut, M., De Beuckeleer, M., and Leemans, J. (1992).Transgenic maize plants by tissue electroporation. *The Plant Cell* 4, 1495–1505.

Fromm, M., Taylor, L.P., and Walbot, V. (1985) Expression of genes transferred into monocot and dicot plant cells by electroporation. *Proc. Natl. Acad. Sci.* USA 82, 5824–5828.

Fromm, M.E., Morrish, F., Armstrong, C., Williams, R., Thomas, J., and Klein, T.M. (1990) Inheritance and expression of chimeric genes in the progeny of transgenic maize plants. *Bio/Technology* 8, 833–839.

Gordon-Kamm, W.J., Spencer, T.M., Mangano, M.L., Adams, T.R., Daines, R.J., Start, W.G., O'Brien, J.V., Chambers, S.A., Adams, W.R., Willetts, N.G., Rice, T.B., Mackey, C.J., Krueger, R.W., Kausch, A.P., and Lemaux, P.G. (1990) Transformation of maize cells and regeneration of fertile transgenic plants. *The Plant Cell* 2, 603–618.

Gould, J., Devey, M., Hasegawa, O., Ulian, E.C., Peterson, G., and Smith, R.H. (1991). Transformation of *Zea mays* L. using *Agrobacterium tumefaciens* and the shoot apex. *Plant Physiol* 95, 426–434.

Grimsley, N.H., Hohn, T., Davies, J.W., and Hohn, B. (1987) *Agrobacterium*-mediated delivery of infectious maize streak virus into maize plants. *Nature* 325, 177–179.

Harris, R., Wright, M., Byrne, M., Varnum, J., Brightwell, B., and Schubert, K. (1988) Callus formation and plantlet regeneration from protoplasts derived from suspension cultures of wheat (*Triticum aestivum* L.) *Plant Cell Reports* 7, 337–340.

He, D.G., Yang, Y.M. and Scott, K.J. (1992) Plant regeneration from protoplasts of wheat (*Triticum aestivum* cv. Hartog). *Plant Cell Reports* 11, 16–19.

Hess, D., Dressler, K., and Nimmrichter, R.(1990) Transformation experiments by pipetting *Agrobacterium* into the spikelets of wheat (*Triticum aestivum* L.). *Plant Science* 72, 233–244.

Klein, T.M., Harper, E.C., Svab, Z., Sanford, J.C., Fromm, M.E., and Maliga, P. (1988) Stable genetic transformation of intact *Nicotiana* cells by the particle bombardment process. *Proc. Natl. Acad. Sci.* USA 85, 8502–8505.

Krens, F.A., Molendijk, L., Wullems, G.J., and Schilperoort, R.A. (1982) In vitro transformation of plant protoplasts with Ti-plasmid DNA. *Nature* 296, 72–74.

Kyozuka, J., Hayashi, Y., and Shimamoto, K. (1987) High frequency plant regeneration from rice protoplasts by novel nurse culture methods. *Mol. Gen. Genet.* 206, 408–413.

Langridge, P., Brettschneider, R., Lazzeri, P., and Loerz, H. (1992) Transformation of cereals via *Agrobacterium* and the pollen pathway: a critical assessment. *The Plant Journal* 2, 631–638.

Larkin, P.J., Taylor, B.H., Gersmann, M., and Brettell, R.I.S. (1990) Direct gene transfer to protoplasts. Aust. J. *Plant Physiol.* 17, 291-302.

Li, L., Qu, R., de Kochko, A., Fauquet, C., and Beachy, R.N. (1993) An improved rice transformation system using the biolistic method. *Plant Cell Reports* 12, 250–255.

Li, Z.Y., Xia, G.M., Chen, H.M., and Guo, G.Q. (1992). Plant regeneration from protoplasts derived from embryogenesis suspension cultures of wheat (*Triticum aestivum* L.) *Journal of Plant Physiology* 139, 714–718.

Luo, Z., and Wu, R. (1988) A simple method for transformation of rice via the pollen-tube pathway. *Plant Mol. Biol.* Reporter 6, 165–174.

Maas, C., and Werr, W. (1989) Mechanism and optimized conditions for PEG mediated DNA transfection into plant protoplasts. *Plant Cell Reports* 8, 148–151.

Mooney, P.A., Goodwin, P.B., Dennis, E.S., and Llewellyn, D.J. (1991) *Agrobacterium tumefaciens*-gene transfer into wheat tissues. *Plant Cell, Tissue and Organ Culture* 25, 209–218.

Nagata, T. (1989) Cell biological aspects of gene delivery into plant protoplasts by electroporation. *International Review of Cytology* 116, 229–255.

Nagata, T., and Takebe, I. (1971) Plating of isolated mesophyll protoplasts on agar medium. *Planta* (Berl.) 99, 12–20.

Paszkowski, J., Shillito, R.D., Saul, M., Mandák, V., Hohn, T., Hohn, B., and Potrykus, I. (1984) Direct gene transfer to plants. EMBO J. 3, 2717–2722.

Potrykus, I. (1990) Gene transfer to cereals: an assessment. *Bio/Technology* 8, 535–542.

Potrykus, I., Saul, M.W., Petruska, J., Paszkowski, J., Shillito, R.D. (1985) Direct gene transfer to cells of a graminaceous monocot. *Mol. Gen. Genet.* 199, 183–188.

Qiao, Y.M., Cattaneo, M., Locatelli, F., and Lupotto, E. (1992) Plant regeneration from long term suspension culture-derived protoplasts of hexploid wheat (*Triticum aestivum* L.). *Plant Cell Reports* 11, 262–265.

Raineri, D.M., Bottino, P., Gordon, M.P., and Nester, E.W. (1990). *Agrobacterium*-mediated transformation of rice (*Oryza sativa* L.) *Bio/Technology* 8, 33–37.

Ren, J.G., Jia, J.F., Li, M.Y., and Zhen, G.C. (1989) Plantlet regeneration from protoplasts isolated from callus cultures of immature inflorescences of wheat (*Triticum aestivum* L.). *Chinese Science Bulletin* 34, 1648–1652.

Rhodes, C.A., Pierce, D.A., Mettler, I.J., Mascarenhas, D., and Detmer, J.J. (1989) Genetically transformed maize plants from protoplasts. *Science* 240, 204–207.

Sanford, J.C., Klein, T.M., Wolf, E.D., and Allen, N. (1987) Delivery of substances into cells and tissues using a particle bombardment process. *Particulate Science and Technology* 5, 27–37.

Shimamoto, K., Terada, R., Izawa, T., and Fujimoto, H. (1989) Fertile rice plants regenerated from transformed protoplasts. *Nature* 338, 274–276.

Somers, D.A., Rines, H.W., Gu, W., Kaeppler, H.F., and Bushnell, W.R. (1992) Fertile, transgenic oat plants. *Bio/Technology* 10, 1589–1594.

Toriyama, K., Arimoto, Y., Uchimiya, H., and Hinata, K. (1988) Transgenic rice plants after direct gene transfer into protoplasts. *Bio/Technology* 6, 1072–1074.

Vasil, V., Redway, F., and Vasil, I.K. (1990) Regeneration of plants from embryogenic suspension culture protoplasts of wheat (*Triticum aestivum* L.). *Bio/Technology* 8,429–433.

Vasil, V., Brown, S.M., Re, D., Fromm, M.E. and Vasil, I.K. (1991). Stably transformed callus lines from microprojectile bombardment of cell suspension cultures of wheat. *Bio/Technology* 9, 743–747.

Vasil, V., Castillo, A.M., Fromm, M.E., and Vasil, I.K. (1992) Herbicide resistant fertile transgenic wheat plants obtained by microprojectile bombardment of regenerable embryogenic callus. *Bio/Technology* 10, 667–674.

Wilmink, A., and Dons, J.J.M. (1993) Selective agents and marker genes for use in transformation of monocotyledonous plants. *Plant Mol. Biol. Reporter* 11, 165–185.

Yamada, Y., Yang, Z.Q., and Tang, D.T. (1986) Plant regeneration from protoplast-derived callus of rice (*Oryza sativa* L.). *Plant Cell Reports* 4, 85–88.

Zhang, H.M., Yang, H., Rech, E.L., Golds, T.J., Davis, A.S., Mulligan, B.J., Cocking, E.C., and Davey, M.R. (1988) Transgenic rice plants produced by electroporation-mediated plasmid uptake into protoplasts. *Plant Cell Reports* 7, 379–384.

GENETIC TRANSFORMATION OF WHEAT

Indra K. Vasil,[1] Vimla Vasil,[1] Vibha Srivastava,[1] Ana M. Castillo,[1] and Michael E. Fromm[2]

[1]Laboratory of Plant Cell and Molecular Biology
1143 Fifield Hall
University of Florida, Gainesville, Florida 32611-0692

[2]Monsanto Company
700 Chesterfield Village Parkway
St. Louis, Missouri 63198

SUMMARY

Transgenic plants of two spring and one winter cultivar of wheat were produced by the direct introduction of DNA, either into long-term regenerable callus cultures, or into immature embryos and young calli. The presence of the herbicide resistant *bar* gene was demonstrated by phosphinothricin acetyltransferase activity, resistance to the herbicide basta, and DNA hybridization. The *bar* gene segregated as a dominant Mendelian trait in sexual progeny.

INTRODUCTION

Agrobacterium tumefaciens is widely used for the efficient genetic transformation of many dicot species. Grass species are outside the host range of this soil-borne bacterium. In spite of extensive efforts and the resultant claims, there is as yet no definitive evidence of stable and heritable transformation of gramineous species by this method. Therefore, many procedures for the direct delivery of DNA to regenerable cells and tissues have been developed to transform grasses. An important prerequisite for the success of this strategy is the efficient regeneration of genetically normal plants from cultured cells. During the past 15 years, the notorious recalcitrance of gramineous species to regenerate plants in vitro has been largely overcome by the judicious use of undifferentiated tissue/organ explants, high concentrations of strong auxins, and the preferential selection of embryogenic cells. Plants have thus been regenerated from protoplast, cell and tissue cultures of all of the major species of grasses, including the economically important cereals (Vasil 1993, Vasil and Vasil 1992,

Improvement of Cereal Quality by Genetic Engineering, Edited by
Robert J. Henry and John A. Ronalds, Plenum Press, New York, 1994

1993). In most instances, plants were derived from somatic embryos, which are formed—directly or indirectly—from single cells.

The rapid advances in the ability to regenerate plants from cultured cells encouraged efforts toward the genetic transformation of grasses. Since the first report of transgenic maize plants in 1988 (Rhodes et al. 1988), fertile transgenic plants resistant to herbicides, insects and viruses have been obtained in all of the important cereal crops (Vasil 1993) by three principal methods of direct DNA delivery: i) polyethylene glycol treatment or electroporation of protoplasts, and ii) high velocity microprojectile bombardment, as well as iii) electroporation, of intact cells and tissues. Wheat was the last of the major cereal crops to be transformed. In this brief report we provide a summary of the work which led to the production of fertile transgenic plants of wheat.

RESULTS AND DISCUSSION

Protocols for the regeneration of wheat plants from embryogenic callus cultures were first established in 1982 (Ozias-Akins and Vasil 1982). These were later extended to regenerable long-term callus cultures, cell suspensions and protoplasts (Redway et al. 1990a,b, Vasil et al. 1990).

Studies on the transient expression of reporter genes delivered into wheat protoplasts showed them to be highly susceptible to damage during electroporation. Since the early attempts to obtain stably transformed cell lines or plants from protoplasts were not successful (Vasil and Vasil, unpublished results), in all of our subsequent work we have used high velocity microprojectile bombardment for the genetic transformation of wheat. A number of genes, including NPTII and EPSPS, were successfully introduced into plated suspension culture cells (Vasil et al. 1991). Although several stably transformed lines were recovered, no plants could be obtained owing to the age of the lines used.

The first transgenic plants of wheat were obtained by delivering pBARGUS plasmid DNA into long-term regenerable type C callus tissue (Vasil et al. 1992). This plasmid contains the selectable *bar* gene, which encodes for the enzyme phosphinothricin acetyltransferase (PAT), and confers resistance to the broad-spectrum herbicide basta by the acetylation of phosphinothricin (PPT), the active ingredient of basta. Callus lines growing in the presence of basta were screened for GUS activity, followed by PAT assays, to confirm the presence of the *bar* gene. Transformed callus tissues were then placed on regeneration media to induce the development of somatic embryos and plants. Integration of the *bar* gene into wheat nuclear DNA was demonstrated by DNA hybridization. The transgenic plants were shown to be resistant to topical applications of basta. Further definitive evidence of transformation was provided by Mendelian segregation of the *bar* gene in sexual progeny.

The type C callus tissue is formed infrequently, and is difficult to identify and maintain in culture. It thus required 12–15 months from the culture of immature embryos to the production of flowering R0 plants. The disadvantages of this system were overcome by delivering DNA directly into cultured immature embryos or young callus tissue (Vasil et al. 1993). Three plasmids (pBARGUS, pMON19526 and pAHC25), each containing the selectable *bar* gene for resistance to the herbicide basta, and the GUS reporter gene, were used. Resistant calli were selected in the presence of PPT, and screened for histochemical GUS activity. Twelve independently transformed callus lines showing PAT activity were recovered from the bombardment of 544 explants, that included 374 immature embryos and 170 one or two month old calli (Table 1). R0 plants were recovered from seven of these lines, five of which have thus far produced R1 progeny. PAT activity was detected in each of the plants tested from the seven R0 lines, as well as in a 1:1 ratio in R1 plants. Resistance to topical application of basta was seen in PAT positive plants and in the transgenic progeny. Molecular analysis by Southern hybridization showed the presence of the *bar* gene in all of the PAT positive R0 and R1 plants analyzed. Hybridization of the *bar* gene probe with high molecular weight DNA further confirmed integration into nuclear DNA. Both male and female transmission of the *bar* gene, and its segregation as a dominant Mendelian trait in R1 plants, were demonstrated (Table 2). Flowering transgenic plants could be obtained in 7–9 months following excision and culture of immature embryos.

Table 1. Summary of six transformation experiments (Vasil et al. 1993)

Expt. # (cultivar)	Age	Plasmid	# Dishes/ explants	PPT+/ GUS+	PAT+ lines	Regen. lines	R0 plants	R1 plants
1 (Pavon)	4d**	pAHC25	5/72	5/2	1P-1	0	0	0
2 (Bob White)	5d**	pAHC25+ pMON19606	4/62	06/4	2B-1	2B-1	10	i.p.
					2B-2	2B-2	10	10
					2B-3	0	0	0
3 (Pavon)	6d**	pAHC25	7/120	12/5	3P-1	3P-1	14	10
4 (Pavon)	9d**	pBARGUS or	8/120	11/4	4P-1	4P-1	25	10#
		pMON19526			4P-2	0	0	0
5 (RH 770019)	1m*	pBARGUS	4/70	6/3	5R-1	5R-1	17	10
					5R-2	0	0	0
					5R-3	0	0	0
6 (RH 770019)	2m*	pAHC25	9/100	10/3	6R-1	6R-1	2	0
6 (Pavon)					6P-1	6P-1	29	10#
Total 6 (3)			37/544	50/21	12	7	7 (107)	5 (50)

Key: * = embryogenic callus, ** = Immature embryos, i.p. = in progress, PPT+ = PPT resistant, # = R2 progeny obtained.

Table 2. Progeny analysis of four independent transgenic lines: 3P-1, 4P-1, 5R-1 and 6P-1 (Vasil et al. 1993)

Transgenic line	# Embryos rescued and progeny analyzed	Basta resistance and/or PAT activity (+/–)
3P-1	10 (cross pollinated)	5/5
4P-1	13 (self pollinated)	9/4
	13 (cross pollinated)	7/6
	35 (cross pollinated)	18/17
5R-1	10 (cross pollinated)	4/6
6P-1	23 (cross pollinated)	11/12

The use of immature embryos and young calli resulted not only in improving the frequency of transformation but also in reducing the time period required to obtain transgenic plants. This should accelerate efforts to further improve this important crop by the introduction of other agronomically useful genes.

Acknowledgements

This work was supported by funds provided to IKV by the Monsanto Company. The plasmid pAHC25 was kindly provided by Dr. Alan H. Christensen.

REFERENCES

Ozias-Akins, P. and Vasil, I.K. (1982) Plant regeneration from cultured immature embryos and inflorescences of *Triticum aestivum* L. (wheat): evidence for somatic embryogenesis. *Protoplasma* 110,95–105.

Redway, F.A., Vasil, V., Lu, D. and Vasil, I.K. (1990a) Identification of callus types for long term maintenance and regeneration from commercial cultivars of wheat (*Triticum aestivum* L.). *Theoret. Appl. Genet.* 79, 609–617.

Redway, F.A., Vasil, V. and Vasil, I.K. (1990b) Characterization and regeneration of wheat (*Triticum aestivum* L.) from embryogenic cell suspension cultures. *Pl. Cell Rep.* 8, 714–717.

Rhodes, C.A., Pierce, D.A., Mettler, I.J., Mascarenhas, D. and Detmer, J.J. (1988) Genetically transformed maize plants from protoplasts. *Science* 240, 204–207.

Vasil, I.K. (1993) Molecular genetic improvement of cereal and grass crops. *IAPTC Newsletter* (in press).

Vasil, I.K. and Vasil, V. (1992) Advances in cereal protoplast research. *Physiol. Plant.* 85, 279 283.

Vasil, I.K. and Vasil, V. (1993) In vitro culture of cereals and grasses. In: *Plant Cell and Tissue Culture*, I.K. Vasil and T.A. Thorpe (eds.), Kluwer Academic Publishers, Dordrecht.

Vasil, V., Brown, S.M., Re, D., Fromm, M.E. and Vasil, I.K. (1991) Stably transformed callus lines from microprojectile bombardment of cell suspension cultures of wheat. *Bio/Technology* 9, 743–747.

Vasil, V., Castillo, A.M., Fromm, M.E. and Vasil, I.K. (1992) Herbicide resistant fertile transgenic wheat plants obtained by microprojectile bombardment of regenerable embryogenic callus. *Bio/Technology* 10, 667–674.

Vasil, V., Redway, F.A. and Vasil, I.K. (1990) Regeneration of plants from embryogenic suspension culture protoplasts of wheat (*Triticum aestivum* L.). *Bio/Technology* 8, 429–433.

Vasil, V., Srivastava, V., Castillo, A.M., Fromm, M.E. and Vasil, I.K. (1993) Rapid production of transgenic wheat plants by direct bombardment of cultured immature embryos. *Bio/Technology* 11, 1553–1558.

APPROACHES TO GENETIC TRANSFORMATION IN CEREALS

K. J. Scott, D. G. He, S. Karunaratne, A. Mouradov, E. Mouradova, and Y. M. Yang

Department of Biochemistry
The University of Queensland
St Lucia 4072
Australia

WHEAT TRANSFORMATION

We describe stable transformation and high frequency transient gus expression of wheat using two approaches, firstly, the electroporation of protoplasts, and secondly, the microprojectile bombardment of immature wheat embryos.

PROTOPLASTS

Embryogenic fine suspension cultures of *Triticum aestivum* cv.Hartog, and *Triticum durum* D6962, established by Yang *et al.* (1991, 1993) were used for the isolation and purification of protoplasts using the methods described by He *et al.* (1992). To examine the viability of the protoplasts before and after electroporation, the protoplasts were stained with fluorescein diacetate (FDA) (Widholm,1972), and the number of fluorescent protoplasts were counted under a fluorescent microscope.

Purified protoplasts were resuspended at a density of 5–10 × 10^6/ml in TBS buffer (Larkin *et al.* 1990), containing 150 mM NaCl, 6 mM $CaCl_2$, 30 mM Tris (pH 9.0) and mannitol to give an osmolality of 0.75 Os/kg. pEmuPAT +/– pEmuGN plasmids (50 µg/ml) were added to the protoplast suspension and the mixture was maintained on ice for 10 min prior to electroporation. The protoplast suspension was transferred to a 1 ml disposable cuvette containing 2 platinum electrodes held 0.6 cm apart. Electroporation was performed using a SEAC electroporator which delivered 10 pulses at 400 v of 3 ms pulse length with 100 ms delay between pulses and using a 120 µF capacitor. Transient expression of the *gus* marker gene was used to establish the optimum parameters for electroporation.

Gus activity was measured by both fluorescence and histological assays .(Jefferson,1987). Following electroporation the protoplasts were maintained on ice for 10 min,

Improvement of Cereal Quality by Genetic Engineering, Edited by
Robert J. Henry and John A. Ronalds, Plenum Press, New York, 1994

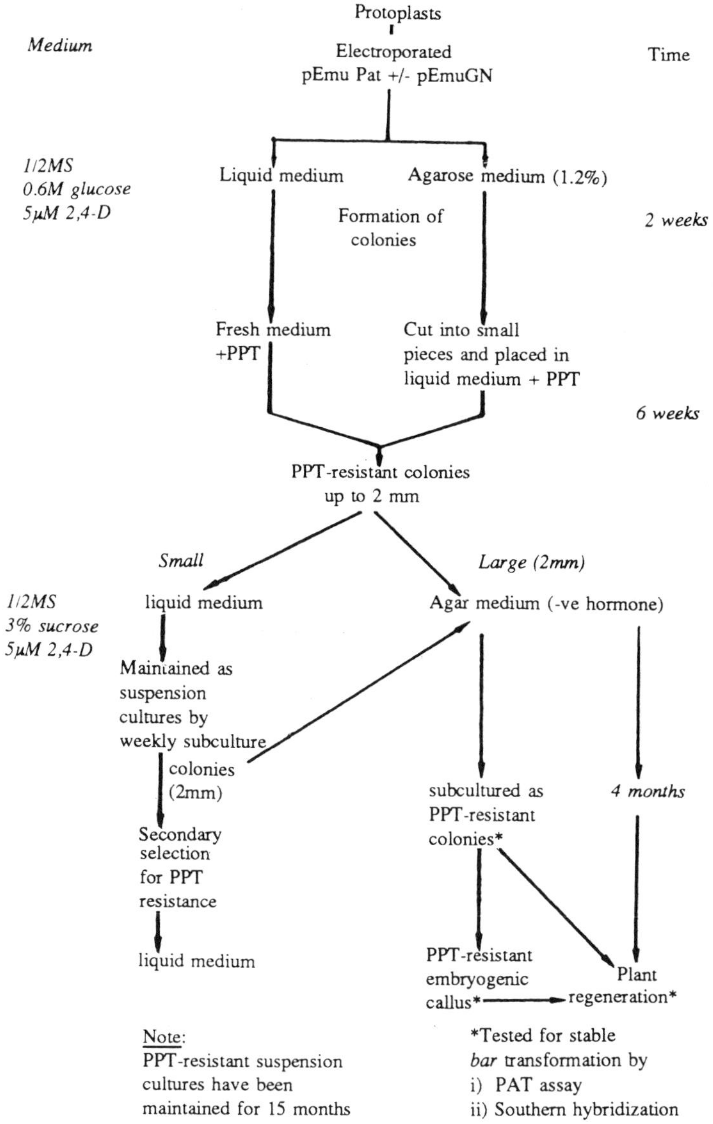

Figure 1. Transformation, selection and regeneration procedures for the production of stably *bar* transformed plants from embryogenic wheat protoplasts.

washed and then transferred to either liquid or solid agarose (1.2%) media. Transformed colonies were selected as outlined in the flow chart (Fig 1).

Southern blot analyses, by hybridization with a PCR fragment comprising 450 bp of the *bar* coding region as a probe, gave positive results for DNA isolated from 19 of 26 randomly selected colonies and callus tissue, including 7 of 8 embryogenic callus tissues (Fig 2). and from leaf material taken from a plant (Fig 3: 3). and from 3 callus tissues which subsequently regenerated into plants (Fig 3: 1, 2, 4). Different multiple bands showed that random integration of the *bar* gene into the wheat genome had occurred. Positive results for phosphinothricin acetyltransferase (PAT) activity (De Block *et al.* 1987) demonstrated the

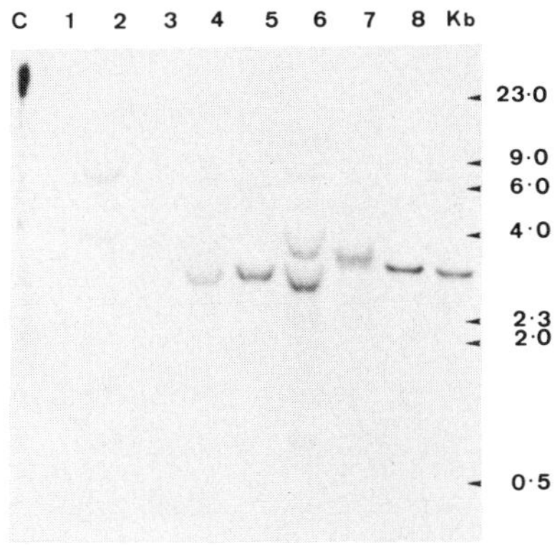

Figure 2. Southern blot analysis of genomic DNA by hybridization with ^{32}P-labelled *bar* coding sequence. DNA samples were digested with *Hin*d III. c: DNA from embryogenic wheat callus (cv. Hartog) regenerated from suspension cells. 1–8: DNA from embryogenic wheat callus (cv. Hartog) regenerated from protoplasts electroporated with pEmuPAT.

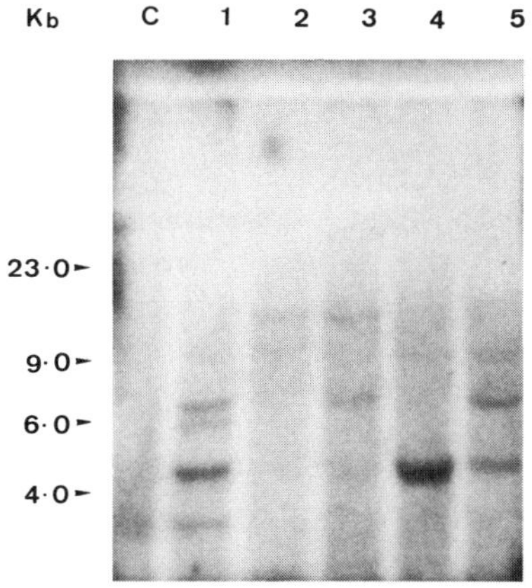

Figure 3. Southern blot analysis of genomic DNA by hybridization with ^{32}P-labelled *bar* coding sequence. DNA samples were digested with *Hin*d III. c: DNA isolated from wheat leaves (cv. Hartog). 1,3,4: DNA isolated from wheat callus tissues (cv. Hartog) which subsequently regenerated into plants. The tissues were regenerated from protoplasts electroporated with pEmuPAT. 2: DNA isolated from wheat leaves (cv. Hartog) taken from a plant which had regenerated from a protoplast electroporated with pEmuPAT. 5: DNA isolated from wheat callus (*T.durum)* regenerated from a protoplast electroporated with pEmuPAT.

presence of the *bar* gene product (Fig 4),). and confirmed the results of Southern blot analysis.

IMMATURE WHEAT EMBRYOS

Triticum aestivum cv. Hartog embryos were harvested from milk-stage embryos and placed on 0.8% agar containing 1/2MS and 2,4-D (2 μg/ml). Two subcultures were performed, at weekly intervals, when all the non-embryogenic tissue associated with each embryo was removed, leaving the highly embryogenic proliferating axis (1–2 mm in width), for use as the target tissue for microprojectile bombardment.

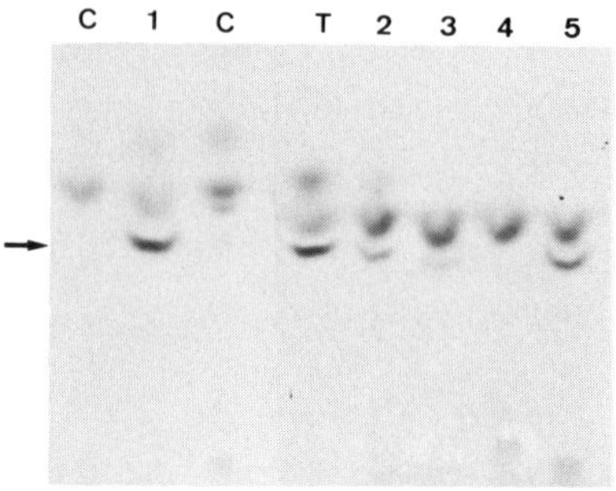

Figure 4. Phosphinothricin acetyltransferase activity. c: embryogenic wheat callus regenerated from suspension cells (cv. Hartog). T: *bar* transformed tobacco leaves. 1–4: wheat callus (cv. Hartog) regenerated from protoplasts electroporated with pEmuPAT. 5: wheat callus (*T. durum*) regenerated from a protoplast electroporated with pEmuPAT →^{14}C-acetylated phosphinothricin.

Target tissues (40–70) were placed on the centre of a filter paper disc (4.2 cm in diameter), dehydrated briefly prior to bombardment with 2 μl suspension of DNA-coated tungsten particles (1.2 μm diameter) using a particle inflow gun (PIG) as described by Finer *et al.* (1992). The bombardment was repeated after the target tissues had been re-orientated on the filter paper. The Tg-DNA suspension was prepared by mixing 5 μl (1 μg/ml) plasmid DNA (100.7-16 PAT) with 25 μl 2.5M $CaCl_2$ and 10 μl 0.1M spermidine (free base). The mixture was vortexed, placed on ice for 5 min prior to the removal of the supernatant (50 μl).

Following bombardment the tissues were incubated on 0.8% agar containing 1/2MS and 2,4-D (2 μg/ml), and tissues, which had been bombarded with tungsten particles without plasmid DNA, were used as non-transformed controls. After 2 weeks any non-embryogenic material was removed from the tissues before they were transferred to fresh medium to which PPT (1 μg/ml) had been added. After a further 2 weeks the cultures were transferred to medium containing 2 μg/ml PPT supplemented with kinetin (2 μg/ml). Following 3–4 weeks the PPT concentration was increased to 3–5 μg/ml and the cultures were exposed to these PPT levels for periods varying from 2–6 weeks. Putative transformed PPT-resistant plants were tested by PCR and leaves were assayed for PAT activity.

After bombardment of the highly embryogenic wheat tissue with 100.7-16 PAT, rapid proliferation and embryogenesis leading to plant regeneration occurred on all tissues in the presence of low levels (1 μg/ml) PPT. Selection of PPT-resistant tissues was performed at varying levels (2–4 μg/ml) PPT. All control tissues died after 2 weeks exposure to 3 μg/ml PPT, whereas approx 10% of the plants, which were regenerated from tissues bombarded with 100.7-17 PAT, continued to grow well in 4–5 μg/ml PPT. DNA, isolated from randomly selected putative *bar* transformed plants, showed the presence of the *bar* gene by PCR analysis and positive results for PAT activity were obtained. Southern blot analysis will be performed when further plant material is available.

IDENTIFICATION AND CHARACTERIZATION OF THE PATHOGEN-INDUCIBLE *CIS*-REGULATORY ELEMENT IN THE BARLEY PR1 GENE

The identification and characterization of the *cis* and *trans* regulatory elements, which control the expression of defence genes, is important to understanding the molecular biology of plant-pathogen interaction.

An *in situ* transient expression assay was designed in our laboratory to analyse the promoter region of the barley PR genes using a histochemical spot-counting assay of transient expression of *gus* activity in the leaves of intact barley plants. This procedure provided a rapid and simple method for the analysis of the promoter region, particularly for cereals where traditional analysis methods are difficult to apply. Using this system we have studied the expression of the barley pathogenesis related (PR) protein gene, *prb1*, following infection with the obligate fungus *Erysiphe graminis* to gain insight into the regulatory mechanisms which govern the expression patterns of the PR-genes and their induction with the fungus. Chimeric vectors carrying deletions of the *prb1* promoter region were fused to the coding region of the *gus* gene. The promoterless *gus* gene was used as the negative control. The positive control was the pEmuGN vector carrying the *gus* gene and showing strong constitutive expression in cereals (Last et al. 1992). Following PIG bombardment of the intact barley plants, which had been grown in Falcon tubes (50 ml), with these chimeric constructs, the leaves of the barley plants were inoculated with *E.graminis*. At 48 and 72 h after inoculation the leaf tissue was stained for transient *gus* activity (Jefferson 1987). Plants which had been bombarded, but not exposed to fungal infection, were used as controls. Following inoculation with the fungus, *in situ* expression of the constructs was estimated by counting the number of blue spots in the bombarded leaves. Using this procedure, the segment of the promoter region containing the regulatory elements was identified.

Acknowledgements

Financial assistance is acknowledged from the Investors in the Queensland Technology Partnership established under the Commonwealth Syndicated R & D program and also from The Alumni Association of the University of Queensland. Acknowledgement is also made to CSIRO Plant Industry, Canberra—pEmuGN and pEmuPAT plasmids bar transformed tobacco (H.Schroeder) Max-Planck Institute, Cologne, Germany—100.7-16 PAT plasmid, Hoechst Australia Ltd—PPT.

REFERENCES

De Block M, Brouwer DD, Tenning P (1989). Transformation of *Brassica napus* and *Brassica oleracea* using *Agrobacterium tumefaciens* and the expression of the *bar* and *neo* genes in the transgenic plants. *Plant Physiol.* 91:694–701.

Finer JJ, Vain P, Jones MW, McMullen MD (1992). Developments of the particle inflow gun for DNA delivery to plant cells. *Plant Cell Reports* 11:323–328.

He DG, Yang YM, Scott KJ (1992). Plant regeneration from protoplasts of wheat (*Triticum aestivum* cv. Hartog). *Plant Cell Reports* 11:16–19.

Jefferson RA (1987). Assaying chimeric genes in plants: the GUS gene fusion system. *Plant Mol Biol Rep* 5:387–405.

Larkin PJ, Taylor BH, Gersmann M, Brettell IS (1990). Direct gene transfer to protoplasts. *Aust J Physiol.* 17:291–302.

Last DI, Brettell RIS, Chamberlain DA, Chaudhury AM, Larkin PJ, Marsh EI, Peacock WJ, Dennis ES (1991). pEmu: an improved promoter for gene expression in cereal cells. *Theor Appl Genet* 81:581–588.

Widholm J (1972). The use of fluorescein diacetate and phenosafranine for determining viability of cultured plant cells. *Stain Technol.* 47:189–194.

Yang YM, He DG, Scott KJ (1991). The establishment of embryogenic liquid cultures and suspension cultures of wheat by continuous callus selection. *Aust J Plant Physiol* 18:445–452.

Yang YM, He DG, Scott KJ (1993). Plant regeneration from protoplasts of durum wheat (*Triticum durum* Desf. cv. D6962). *Plant Cell Rep* 12:320–323.

GENETIC ENGINEERING OF WHEAT AND BARLEY*

K. K. Kartha,† N. S. Nehra, and R. N. Chibbar

Plant Biotechnology Institute
National Research Council
110 Gymnasium Place
Saskatoon, Saskatchewan
Canada, S7N 0W9

SUMMARY

We have obtained stably transformed cell lines of wheat and barley by biolistics-mediated delivery of an improved plasmid pRC62 (ACT-1D-GUS::NPTII-NOS) into the non-embryogenic cell cultures. Enzyme assays and Southern hybridization confirmed functional expression and stable integration of the chimeric genes into the wheat and barley genomes. These results demonstrated that efficient delivery and stable expression of foreign genes into wheat and barley cells was no longer a constraint for accomplishing genetic engineering of these crops. However, the major problem which remained was the lack of an efficient system for regeneration of fertile plants from transformed cells. We have now resolved this problem by developing a system for enhanced somatic embryogenesis and plant regeneration from isolated scutellar tissue of wheat and barley. This scutellar-based regeneration system has been successfully used for developing a simple and reproducible protocol for the production of self-fertile transgenic wheat plants. Using this protocol, transgenic wheat plants carrying the *bar*, *gus* and *npt* genes have been produced. Molecular and biochemical analyses confirmed stable integration and functional expression of transgenes in R_0 and R_1 transgenic plants. Mendelian inheritance of *bar* gene was observed in R_1 and R_2 progeny. The protocol is being optimized for production of transgenic barley plants. Since the protocol avoids the need for establishing long-term callus, cell suspension or protoplast regeneration procedures, it has the potential for becoming a practical system for gene transfer in wheat and barley. The procedure will facilitate manipulation of various agronomic and quality traits by direct gene transfer into these major cereal grain crops.

---

*NRCC Publication No. 37336.

†Corresponding author.

Improvement of Cereal Quality by Genetic Engineering, Edited by
Robert J. Henry and John A. Ronalds, Plenum Press, New York, 1994

INTRODUCTION

Cereal grain crops have held a prominent place among the food crops of the world ever since the first domestication of cultivated plants by humankind. The important cereal grain crops such as wheat, rice, corn, barley, oats, rye, sorghum, and millets have been regarded as a principal source of carbohydrates, proteins, and minerals in human and animal diet. In addition, cereals are also an important source of starch that can be converted into ethanol for use in beverages and as a source of fuel. Among the cereal grain crops, wheat and barley together account for more than 40 percent of the total world grain production (Agri-Food Perspectives, 1992) and thus occupy a unique position in the global agricultural economy. As a result these crops have been a target for intensive and extensive research activities with prime emphasis on the use of plant breeding techniques for improvement of yield, quality and other agronomic traits. Although conventional plant breeding has made tremendous contributions to the development of improved high-yielding varieties of both wheat and barley, the efforts of classical breeders are often limited by the availability of a restricted gene pool of these species and the inability to manipulate traits in a directed manner. With the recent advancements made in the field of plant molecular biology, there is every reason to be optimistic that further improvement of the existing superior varieties of wheat and barley by incorporation of unique value-added traits will increasingly depend on the modern biotechnological approach of direct gene transfer.

The successful employment of biotechnological approaches for crop improvement requires a simple and reproducible genetic transformation procedure. In recent past, considerable efforts have been made around the world for developing a transformation technology for wheat and barley. However, the progress made in this direction has been slow mainly because of the recalcitrance of these species to *Agrobacterium*-mediated gene insertion, difficulties encountered in the establishment of embryogenic cell cultures or protoplast regeneration methods capable of producing fertile plants, and lack of efficient DNA delivery systems. It is evident from the review of progress made in transformation technology for cereals that some of these impediments have been overcome resulting in production of transgenic plants for cereals such as rice (Shimamoto et al., 1989; Datta et al., 1990), maize (Fromm et al., 1990; Gordon-Kamm et al., 1990), wheat (Vasil et al., 1992), and oats (Somers et al., 1992). However, in most cases the production of transgenic plants was dependent upon the establishment of long-term callus or an embryogenic cell culture or a protoplast regeneration system which obviously limited the use of transformation protocols to only those genotypes or inbred lines that were amenable to such *in vitro* manipulations. In addition, such procedures were not only time consuming but also resulted in production of a large proportion of infertile plants among regenerants probably as a consequence of tissue culture-induced variation. As an alternative to cell culture and protoplast mediated transformation procedures, the use of immature zygotic embryos as a gene recipient system has been proposed (Kartha et al., 1989; Chibbar et al., 1991), which has recently resulted in the production of transgenic rice (Christou et al., 1991), maize (D'Halluin et al., 1992) and wheat (Nehra et al., 1993). The development of such transformation procedures has implications for genetic modification of a wide variety of commercial genotypes.

In this report we describe the production of transgenic cell lines of wheat and barley from non-embryogenic cell cultures using the enhanced gene expression vectors and efficient selection strategies (Chibbar et al., 1993; Chibbar et al., unpublished). We also report on the development of an enhanced regeneration system (ERS) from isolated scutellar tissues of wheat and barley. The system was successfully used for the development of a simple and reproducible procedure for incorporation of two distinct gene construct into wheat genome, and the production of self-fertile herbicide resistant transgenic wheat plants (Nehra et al., 1993). This system has potential for becoming a practical method for gene transfer into wheat, barley and possibly other cereal grain crops.

MATERIALS AND METHODS

Plant Material

Plants of different wheat (*Triticum aestivum* L.) and barley (*Hordeum vulgare* L.) varieties were grown at regular intervals, in 6 inch diameter plastic pots filled with Ready Earth potting mixture, to maintain a steady supply of plant material. Plants were maintained in a growth chamber at 25 ± 2°C day and 20 ± 2°C night temperatures under a 16 h photoperiod (150 $\mu E.m^{-2}s^{-1}$) provided by banks of fluorescent tubes and incandescent bulbs. All plants were watered every second day and fertilized once a week with 0.4 g/l of soluble greenhouse fertilizer (20:20:20).

Establishment and Maintenance of Cell Suspension Cultures

Non-embryogenic cell suspension cultures of wheat (var. HY320) and barley (var. Heartland) used in this study were established from immature zygotic-embryo-derived friable calli according to procedures described in previous papers (Kartha et al., 1989; Chibbar et al., 1993). The fast growing cell suspension cultures were maintained by subculturing, twice a week, 15–20 ml of suspension into 50 ml of fresh liquid MS (Murashige and Skoog, 1962) medium supplemented with 3% sucrose and 1.1 mg/l of 2,4 dichlorophenoxy aceticacid (2,4-D). The medium was solidified with 0.8% Difco-Bacto agar when needed. For selection experiments, a filter-sterilized solution of Geneticin (G418 sulphate) was added to the liquid or solid medium as required to obtain desired concentrations.

Preparation and Culture of Scutella

Wheat and barley spikes (8 to 12 days-post anthesis) were harvested from greenhouse-grown plants. Immature caryopses were removed from the spikelets and surface sterilized with 70% ethanol for one minute followed by a treatment with 20% (v/v) Javex bleach (1.2% sodium hypochlorite) for 20 minutes for wheat or 5% (v/v) Javex bleach for 5 minutes for barley, and rinsed 5 times with sterile double distilled water. Using a stereo dissecting microscope, the immature embryos were then aseptically removed from the caryopses and the scutella were isolated by carefully removing the entire embryo axis. The isolated scutella were cultured according to the conditions described earlier (Nehra et al., 1993).

Gene Constructs

The plasmid pBARGUS (Fromm et al., 1990) used in this study contains a *bar* gene expressing under the control of cauliflower mosaic virus 35S (CaMV 35S) promoter and alcohol dehydrogenase 1 (Adh1) intron 1 at 5′ region and a nopaline synthase (nos) terminator at 3′ end. It also contains a *gus* marker gene driven by Adh1 promoter and its first intron and a nos terminator. The plasmid pRC62 contains a GUS::NPTII fusion gene (Datla et al., 1991) driven by rice actin promoter and its first intron (McElroy et al., 1990) and a nos terminator.

DNA Coating and Bombardment

Gold (1.0 μm) or tungsten (M 10) particles were coated with plasmid DNA using the procedure described previously (Chibbar et al., 1993). Actively growing cell suspension cultures were prepared for particle bombardment essentially according to Kartha et al. (1989). The procedures for preparation and bombardment of isolated scutella were same as described earlier (Nehra et al., 1993).

Selection of Transformants

The filter papers carrying cells bombarded with plasmid DNA were cultured for two days on agar-solidified MS medium and then selected for four weeks by floating the filter papers every week on fresh liquid medium containing increasing levels of Geneticin (10 to 30 mg/l). The cells were subsequently selected on 40 mg/l Geneticin for 4 weeks by weekly transfer on fresh liquid medium. Following 8 weeks of selection, the actively growing colonies were removed and maintained as independent callus lines for further growth and analysis. The bombarded scutella were selected according to selection strategies described previously (Nehra et al., 1993) using L-phosphinothricin (L-PPT) and Geneticin as selection agents. The selected shoots were transferred to half MS for further growth and rooting. The plantlets developed on this medium were transferred to environmentally controlled growth chambers and grown for further analysis.

Enzyme Assays and Herbicide Testing

Histochemical and fluorometric assays for β-glucuronidase (GUS) activity were performed as described earlier (Chibbar et al., 1991). Neomycin phosphotransferase (NPTII) activity was detected using the dot-blot assay procedures as described in a previous report (Nehra et al., 1990) except that leaf tissues were extracted using GUS lysis buffer of Jefferson et al. (1987). Phosphinothricin acetyltransferase (PAT) assay was performed using the silica gel thin layer chromatography procedure according to Spencer et al. (1990). Herbicide testing of transgenic plants was carried out by either brushing the young leaves of control and transgenic plants with an aqueous (50%) solution of L-PPT (the active ingredient of commercial herbicide Basta or Ignite) at different concentrations (0.004-0.5%, v/v) containing 0.1% (v/v) Tween 80 (Sigma) or by spraying the young plants with 0.3 % (v/v) solution of Ignite (Basta). The L-PPT solution was applied with a paint brush to both sides of leaves. Plants were assessed for damage one week after herbicide application.

Plant DNA Isolation and Southern Blot Hybridization

Total genomic DNA was extracted from freeze dried calli or young leaf tissues (0.5 g) of primary transformants and progeny plants according to the procedure described by Doyle and Doyle (1990). Undigested or DNA digested by restriction enzymes using the conditions recommended by the suppliers (Boehringer) was electrophoresed on a 1% agarose gel. The DNA was transferred from the agarose gel to nylon membranes (HyBond) which were prehybridized and subsequently hybridized with radiolabelled probes according to standard procedures (Ausubel et al., 1989). The probes were labelled with ^{32}P by random primed DNA labelling (Promega) kit. The hybridized filters were washed 3 times at 42°C with 0.1% SSC and 0.1% SDS and analyzed by autoradiography.

RESULTS AND DISCUSSION

A. Stable Transformation of Wheat and Barley Cell Cultures

Bombardment and Selection of callus lines. Our previous studies with barley cell suspension cultures which were designed to test the efficacy of various promoter and intron combinations showed that rice actin promoter with its first intron gave the highest level of transient *gus* expression (Chibbar et al., 1993). Antibiotic sensitivity tests with cell suspension cultures or immature zygotic embryos of wheat and barley showed that geneticin (G418 sulphate) was more effective selection agent than hygromycin. The selection experiments with cell suspension cultures were therefore undertaken to test the efficacy of plasmid pRC62 (ACT-1D-GUS::NPTII-NOS) for production of stably transformed cell lines of wheat and barley using geneticin as a selection agent. The results from these experiments proved that rice actin promoter and geneticin selection was very effective in producing transformed cell lines of both wheat and barley. A total of 37 geneticin resistant colonies were obtained from 24 filters of wheat cells bombarded with pRC62. In similar bombardment experiments with barley cells 3 independent geneticin resistant colonies have so far been obtained from different experiments. The selected callus lines of both wheat and barley grew vigorously on solid MS medium containing 40 mg/l geneticin whereas the non-transformed control callus failed to show any growth at this concentration (Fig. 1A).

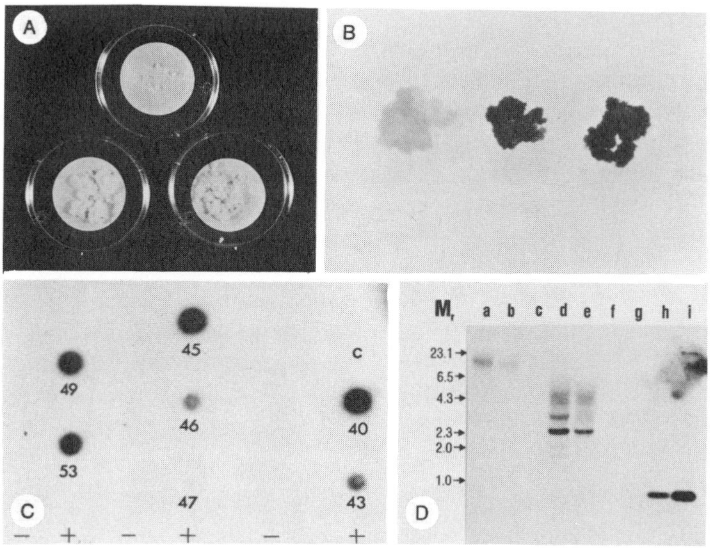

Figure 1. Stable transformation of wheat cell cultures. **(A)** Growth of a non-transformed control (top) and two independently transformed callus lines of wheat (bottom) on 40 mg/l geneticin 4 weeks after selection. **(B)** Expression of GUS activity in control (left) and two transgenic callus lines of wheat (right). **(C)** Analysis of NPTII activity in control (c) and several transgenic callus lines of wheat. For each extract the reaction was carried out with (+) or without (−) neomycin in the reaction mixture. **(D)** Southern blot analysis of DNA from transgenic callus lines. Lanes: undigested genomic DNA from transgenic callus lines (a & b); uncut (c) and DNA cut with Xba1 and EcoR1 (f & g) from non-transformed control callus; DNA from transgenic callus lines digested with Xba1 and EcoR1 (d & e); five and ten pg of 1 kb *npt* fragment used as positive control (h & i). The membrane was hybridized with a radiolabelled *npt* probe. Mr shows the position of molecular wt. markers in kb.

Enzyme and Molecular Analysis of Selected Callus Lines. The selected callus lines of wheat and barley were assayed histochemically for the GUS enzyme activity by transferring a small portion of callus into X-gluc solution. All geneticin resistant callus lines showed a uniform blue colour that confirmed the presence of a functional GUS enzyme in transformed callus lines (Fig. 1B). Some of the GUS-positive callus lines were randomly chosen for the analysis of NPTII activity using the dot-blot assay. All GUS-positive callus lines tested showed varying levels of NPTII enzyme activity (Fig.1C) indicating the presence of an intact GUS::NPTII expression unit in the transformed callus lines. Total genomic DNA isolated from several transformed callus lines of wheat and barley was hybridized to *gus* and *nptII* probes to confirm the integration of plasmid DNA into host genome. The results of Southern hybridization of two transgenic callus lines of wheat are shown in Fig. 1D. The hybridization of undigested genomic DNA with *nptII* probe showed integration of fusion gene in the high molecular weight DNA of wheat. The genomic DNA digested with XbaI and EcoRI showed a 3 kb hybridization fragment indicating the presence of entire *gus::nptII* fusion gene in the transformed callus lines.

B. Production of Fertile Transgenic Wheat Using an Enhanced Regeneration System

Development of an Enhanced Regeneration System. A novel system for enhanced somatic embryogenesis from isolated scutellar tissue of wheat and barley has been developed. *In vitro* culture of isolated scutella of wheat var. Fielder resulted in high frequency (85–100%) somatic embryogenesis with a propensity to produce a large number of distinct somatic embryos (10–15) from a single explant within 2–3 weeks from initiation of cultures. In this system, the onset of somatic embryogenesis was marked by the appearance of a transparent circle on the adaxial (concave) surface in the proximal half of isolated scutella within 2–3 days after initiation of cultures. The appearance of this circle was an indication of early cell division activity of competent embryogenic cells contained in the scutella which gave rise to a mass of nodular creamy embryogenic callus, consisting of globular stage somatic embryos, within a week from the initiation of cultures. Transfer of cultures at this stage to low light (10 $\mu E.m^{-2}s^{-1}$) resulted in conversion of embryogenic callus into a cluster of distinct somatic embryos which further developed into mature somatic embryos (Fig. 2A) in about two weeks. The process of somatic embryogenesis in barley was similar to wheat except that the embryogenic potential was not confined to a particular zone of scutellum. In barley the entire surface of scutellum turned transparent and formed globular embryos within a week which further developed into mature somatic embryos (Fig. 2B) within 3 weeks from initiation of cultures. A large proportion (50–70%) of mature somatic embryos of both wheat and barley were easily converted into complete plantlets on half-strength. hormone-free MS

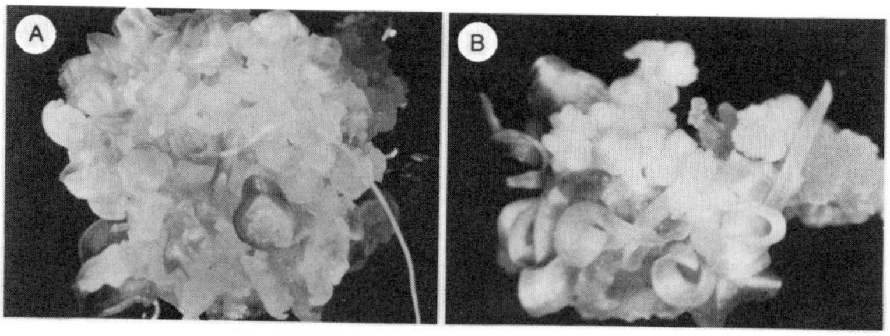

Figure 2. Enhanced somatic embryogenesis from isolated scutella of wheat and barley. Mature somatic embryos developed on isolated scutellum of wheat (**A**) and Barley (**B**) three weeks after culture initiation.

medium in another two weeks. The rooted plantlets were successfully established in a greenhouse and grown into phenotypically normal and fully fertile plants. The system was applicable to several commercial varieties of wheat and barley. However, for the purpose of developing a reproducible transformation system, we focused our efforts on a commercial wheat var. Fielder.

Selection of Transformants. The isolated scutella of var. Fielder bombarded with pBARGUS were allowed to form somatic embryos which were then selected in the presence of L-phosphinothricin (L-PPT), an active ingredient of commercial herbicide Basta or Ignite. Although some clusters of non-transformed somatic embryos remained green after 4 weeks of selection, their growth was completely arrested at 5 mg/l L-PPT. However, the putatively transformed somatic embryos developed into 2–3 cm long green shoots after 4 weeks of selection on 5 mg/l L-PPT. In six different transformation experiments performed with pBARGUS using L-PPT as a selective agent, 16 putative transformants were recovered from a total of 942 scutella bombarded. In a second set of experiments the isolated scutella were bombarded with pRC62 (ACT-1D-GUS::NPTII-NOS). Following bombardment, the intact scutella were selected on 30 mg/l geneticin for 4 weeks to enrich the growth of transformed cells and the formation of somatic embryos. The somatic embryos formed on this medium were then selected on 75 mg/l geneticin for obtaining putative transformants. The results of transformation experiments using this two-step selection strategy, employing geneticin as a selective agent, yielded 19 putative transformants from 950 scutella bombarded. The regenerants obtained from selection experiments were transferred to hormone-free medium for development into plantlets. All the plantlets were successfully established in the greenhouse and grown into fully fertile mature plants.

Analyses of Primary Transformants. The expression of the *bar* gene product in R_0 plants transformed with pBARGUS was assessed by analysis of phosphinothricin acetyltransferase (PAT) activity in young leaf tissues obtained from greenhouse grown plants. Out of 16 plants selected on L-PPT, 8 showed varying levels of PAT activity. However, none of the plants selected on L-PPT showed GUS activity in histochemical and fluorometric GUS assays. The young leaves of PAT positive plants when challenged with L-PPT (0.004–0.4%) showed varying levels of tolerance to herbicide. A strong correlation of PAT enzyme activity with the level of herbicide tolerance was observed in R_0 transformants. The total genomic DNA isolated from leaf tissues of PAT-positive primary transformants was hybridized with *bar* and *gus* probes to confirm the integration of plasmid DNA into wheat chromosomes. The hybridization of uncut genomic DNA of four primary transformants with *bar* probe showed integration of the *bar* gene in the high molecular weight DNA of wheat. Southern analysis of DNA digested with different restriction enzymes confirmed the presence of complete BAR and GUS expression cassettes in these transformants. In addition, this analysis also provided evidence for integration of an incomplete GUS expression cassette in some of the transformants.

Histochemical and fluorometric GUS assays were performed on leaf pieces taken from primary transformants obtained with pRC62. Out of 19 plantlets selected on geneticin, 15 were positive for GUS activity. The GUS-positive plants also exhibited varying levels of NPTII activity in dot-blot enzyme assays indicating the functional expression of both *GUS* and NPTII genes in transgenic wheat plants. Southern blot analysis of GUS and NPTII positive plants confirmed the integration of introduced plasmid in high molecular weight genomic DNA of wheat and also indicated the presence of entire GUS::NPTII fragment in transformants.

Analyses of Progeny Plants. The primary transformants obtained from both gene constructs were grown to maturity in the greenhouse. All the transformants developed into normal and fully self-fertile plants. Selfing of R_0 transformants resulted in normal seed set and produced viable seeds from each individual tiller. Mature seeds collected from two independent tillers of some R_0 transformant were germinated to obtain R_1 progeny. Segregation of the *bar* gene in one-month-old progeny was determined by spraying the plants with 0.3% (v/v) solution of Ignite (Basta). The results of segregation analysis based on herbicide resistance in leaves confirmed the stable transmission and Mendelian inheritance of *bar* gene in the progeny of independent transformants (Fig. 3A). The results of herbicide testing were also supported by the segregation of PAT activity in R_1 progeny plants. Although *gus* was not expressed in progeny plants, Southern blot analysis of R_1 plants (3 PAT-positive and 1 PAT-negative) obtained from two independent transformants revealed the presence of both *bar* and *gus* coding sequences in the plants expressing PAT activity and herbicide resistance. The segregation of *bar* gene was also observed in R_2 progeny plants where it was possible to distinguish homozygous and heterozygous progeny plants (Fig. 3B). The seed collected from homozygous lines is being multiplied for further analysis. Histochemical GUS staining of leaf pieces obtained from R_1 progeny plants transformed with pRC62 also showed that the GUS::NPTII fusion gene was inherited in the progeny.

Figure 3. Segregation analysis of herbicide resistance in transgenic wheat. **(A)** A control (left) and R_1 progeny plants (right) of a transgenic wheat line showing 3:1 segregation of *bar* gene. **(B)** Non-transformed control plants (centre) surrounded by plants of from a herbicide resistant homozygous line in R_2 progeny of transgenic wheat.

Conclusions and Prospects

In this study we have demonstrated the development of an enhanced regeneration system for high frequency somatic embryogenesis from isolated scutellar tissue of wheat and barley. We have also developed improved monocot gene expression vectors and efficient selection strategies for the production of stably transformed cell lines of wheat and barely. By combining these parameters with biolistics-mediated DNA delivery into competent embryogenic cells of scutellar tissue, we have succeeded in developing a simple and reproducible procedure for genetic transformation of wheat. Using the transformation procedure described in this study, transgenic plants of wheat ready for transfer into a greenhouse were recovered from different experiments at frequencies in the range of 0.5 to 2.5 percent within 3 months from initiation of cultures. We have used the system for successful incorporation of two distinct gene constructs, carrying *bar, gus* and GUS::NPTII fusion gene driven by three different promoters, into wheat genome. Additionally, all the primary transformants regenerated in our study were phenotypically normal and fully fertile. Furthermore, our results demonstrate that transgenes were stably transmitted to R_1 and R_2 progeny plants in a Mendelian fashion. These results conclusively demonstrate that genetic transformation of wheat has been accomplished by particle bombardment of isolated scutellar tissue. Although the results are described in this study for only one genotype of wheat, the system can be easily adapted for other genotypes of wheat, barely and possibly other cereal grain crops. We are currently working on extending this technology to various commercial genotypes of wheat and barley. The development of this technology now opens up the possibilities for direct manipulations of cereals for improved agronomic performance and grain quality. Our efforts are also directed at improving the quality of starch in wheat by modifying the genes involved in starch biosynthesis.

Acknowledgements

We wish to thank Dr. M. Fromm for providing the pBARGUS; Drs. D. McElroy and R. Wu for the rice actin promoter; and Drs. W. Crosby and R. Datla for the GUS::NPTII fusion gene. We are also thankful to Nick Leung, Karen Caswell, Lee Steinhauer and Cliff Mallard for their expert technical help. This research was partially supported by Western Grains Research Foundation of Canada through the award of a research grant.

REFERENCES

Agri-Food Perspectives. (1992). World Supply and Disposition of Wheat. (Agriculture Canada: Economic Analysis Division).

Ausubel, F.M., Brent, R., Kingston, R.E., Moore, D.D., Seidman, J.G., Smith, J.A., and Struhl, K., eds (1990). *Current Protocols in Molecular Biology.* (New York: Greene Publishing Associates / Wiley-Interscience).

Chibbar, R.N., Kartha, K.K., Leung, N., Qureshi, J., and Caswell, K. (1991). Transient expression of marker genes in immature zygotic embryos of spring wheat (*Triticum aestivum)* through microprojectile bombardment. *Genome* 34, 453–460.

Chibbar, R.N., Kartha, K.K., Datla, R.S.S., Leung, N., Caswell, K., Mallard, C., and Steinhauer, L. (1993). The effect of different promoter-sequences on transient expression of *gus* reporter gene in cultured barley (*Hordeum vulgare* L.) cells. *Plant Cell Rep.* 12, 506–509.

Christou, P., Ford, T.L., and Kofron, M. (1991). Production of transgenic rice (*Oryza sativa* L.) plants from agronomically important Indica and Japonica varieties via electric discharge particle acceleration of exogenous DNA into immature zygotic embryos. *Bio/Technol.* 9, 957–962.

Datla, R.S.S., Hammerlindl J., Pelcher, L.E., Crosby, W.L., and Selvaraj, G. (1991). A bifunctional fusion between β-glucuronidase and neomycin phosphotransferase: a broad-spectrum marker enzyme for plants. *Gene 101*, 239–246.

Datta, S.K., Peterhans, A., Datta, K., and Potrykus, I. (1990). Genetically engineered fertile Indica-rice recovered from protoplasts. *Bio/Technol.* 8, 736–740.

D'Halluin, K., Bonne, E., Bossut, M., Beuckeleer, M., and Leemans, J. (1992). Transgenic maize plants by tissue electroporation. *Plant Cell* 4, 1495–1505.

Doyle, J.J., and Doyle, J.L. (1990). Isolation of plant DNA from fresh tissue. *Focus* 12, 13–15.

Fromm, M.E., Morrish, F., Armstrong, C., Williams, R., Thomas, J., and Klein, T.M. (1990). Inheritance and expression of chimeric genes in the progeny of transgenic maize plants. *Bio/Technol.* 8, 833–839.

Gordon-Kamm, W. J., Spencer, T.M., Mangano, M.L., Adams, T.R., Daines, R.J., Start, W.G., O'Brien, J.V., Chambers, S.A., Adams, W.R., Willetts, N.G., Rice, T.B., Mackey, C.J., Krueger, R.W., Kausch, A.P., and Lemaux, P.G. (1990). Transformation of maize cells and regeneration of fertile transgenic plants. *Plant Cell* 2, 603–618.

Jefferson, R.A., Kavanagh, T.A., and Bevan, M.W. (1987). GUS fusion: β-Glucuronidase as a sensitive and versatile gene fusion marker in higher plants. *EMBO* J. 6, 3901–3907.

Kartha, K.K., Chibbar, R.N., Georges, F., Leung, N., Caswell, K., Kendall, E., and Qureshi, J. (1989). Transient expression of chloramphenicol acetyltransferase (CAT) gene in barley cell cultures and immature embryos through microprojectile bombardment. *Plant Cell Rep.* 8, 429–432.

McElroy, D., Zhang W., Cao, J., and Wu, R. (1990). Isolation of an efficient actin promoter for use in rice transformation. *Plant Cell* 2, 163–171.

Murashige, T., and Skoog, F. (1962). A revised medium for rapid growth and bioassays with tobacco tissue cultures. *Physiol. Plant.* 415, 473–497.

Nehra, N.S., Chibbar, R.N., Kartha, K.K., Datla, R.S.S., Crosby, W.L., and Stushnoff, C. (1990). Genetic transformation of strawberry by *Agrobacterium tumefaciens* using a leaf disk regeneration system. *Plant Cell Rep.* 9, 293–298.

Nehra, N.S., Chibbar, R.N., Leung, N., Caswell, K., Mallard, C., Steinhauer, L., Baga, M., and Kartha, K.K. (1993). Self-fertile transgenic wheat plants regenerated from isolated scutellar tissues following microprojectile bombardment with two distinct gene constructs. *Plant J.* (Submitted).

Shimamoto, K., Terada, R., Izawa, R., and Fujimoto, H. (1989). Fertile transgenic rice plants regenerated from transformed protoplasts. *Nature* 338, 274–276.

Somers, D.A., Rines, H.W., Gu, W., Kaeppler, H.F., and Bushnell, W.R. (1992). Fertile, transgenic oat plants. *Bio/Technol.* 10, 1589–1594.

Spencer, T.M., Gordon-Kamm, W.J., Daines, R.J., Start, W.G., Lemaux, P.G. (1990). Bialaphos selection of stable transformants from maize cell culture. *Theor. Appl. Genet.* 79, 625–631.

Vasil, V., Castillo, A.M., Fromm, M.E., Vasil, I.K. (1992). Herbicide resistant fertile transgenic wheat plants obtained by microprojectile bombardment of regenerable embryogenic callus. *Bio/Technol.* 10, 667–674.

GENETIC ENGINEERING IN RICE PLANTS

Hirofumi Uchimiya[1] and Seiichi Toki[2]

[1]Institute of Molecular and Cellular Biosciences
University of Tokyo
Bunkyo-ku, Tokyo l 13, Japan

[2]Department of Biological Science
Faculty of Science
Hokkaido University
Sapporo 060, Japan

SUMMARY

Recently, gene transfer into higher plants has made it possible to analyse foreign gene expression in transgenic rice plants. We have constructed a chimeric gene consisting of the promoter, 1st exon, and lst intron of a maize polyubiquitin gene (Ubi-l) and the coding sequence of the *bar* gene from *Streptomyces hygroscopicus.* This construct was transferred into rice protoplasts via electroporation. Transgenic plants grown in a greenhouse were resistant to both bialaphos and phosphinotliricine at a dosage lethal to untransformed control plants. Western blot analysis and enzymatic assays verified expression of the active bar gene-product. Apparent Mendelian segregation for bialaphos resistance and enzymatic activity in T1 progeny of primary transformants are consistent with heritable transmission of the introduced marker gene. When rice plants expressing a bar gene under the control of the maize polyubiquitin promoter were subjected to mycelium of the sheath blight disease pathogen, *Rhizoctonia solani,* followed by bialaphos treatment, only transgenic plants survived and did not show the disease symptoms. Thus, bialaphos resistant rice plants could be useful for prevention of fungal pathogen attack.

INTRODUCTION

The genetic improvement in rice has been achieved mainly through the application of classical Mendelian genetics and conventional plant breeding methods. Recent advance in biotechnology particularly in cell culture and molecular biology has opened new avenues in the genetic improvement of crop plants. Some of the exciting developments in rice biotech-

Improvement of Cereal Quality by Genetic Engineering, Edited by
Robert J. Henry and John A. Ronalds, Plenum Press, New York, 1994

nology include regeneration of plants from protoplasts of both indica and japonica species, production of transgenic plants, and development of molecular (RFLP) map. These advances have widened the scope and potentials of DNA transformation in rice.

Although the production of transgenic rice plants has been reported by a number of laboratories (Battraw and Hall, 1990; Christou *et al.*, 1991; Datta *et al.*, 1990; Hayashimoto *et al.*, 1990; Peng *et al.*, 1990; Shimamoto *et al.*, 1989; Tada *et al.*, 1990; Terada and Shimamoto, 1990; Toriyama *et al.*, 1988; Zhang *et al.*, 1988), the regeneration of fertile, stably transformed rice is yet to be made routine and efficient. In many plant transformation studies, drug resistance genes such as neomycin phosphotransferase or hygrornycin phosphotransferase have been used as selectable markers (Shimamoto *et al.*, 1989; Tada *et al.*, 1990; Toriyama *et al.*, 1988; Tada *et al.*, 1991). One recent alternative strategy that has emerged is based on the use of marker genes that confer resistance to herbicides (De Block *et al.*, 1987; De Block *et al.*, 1989; Fromm *et al.*, 1990; Gordon-Kamm *et al.*, 1990). Bialaphos and phosphinothricin (PPT) are known to be potent herbicides with a broad spectrum of toxicity to numerous crops as well as weeds. Bialaphos was first registered as a tripeptide antibiotic produced by *Streptomyces hygroscopicus* SF1293. It consists of Phosphinothicin (PPT), an analogue of L-glutamic acid, and two L-alanine residues. Upon removal of the alanine residues by endogenous peptidases in plant cells, the resulting PPT inhibits glutamine synthetase, thus causing a rapid accumulation of ammonia that leads to plant cell death (Tachibana *et al.*, 1986a; 1986b). The *bar* gene cloned from *Streptomyccs hygroscopicus,* encodes the enzyme phosphinothricin acetyltfansferase (PAT) which acetylates the NH_2 terminal group of PPT abolishing its herbicidal activity (Murakami *et al.*, 1986; Tompson *et al.*, 1987). Chimeric constructs consisting of the CaMV 355 promoter fused to the bar gene have been transferred into tobacco, tomato, and potato (De Block *et al.*, 1987) and rape (De Block *et al.*, 1989) through Agrobacterium-mediated transformation, and such transgenic plants were resistant to both PPT and bialaphos. Similarly, the same gene has been introduced into the monocot plant, maize, by micro-projectile bombardment with the production of fertile, transgenic plants (Fromm *et al.*, 1990; Gordon-Kamm *et al.*, 1990). In the case of rice (*Oryza sativa* L.), Dekeyser *et al.* (1989) pointed out the potential usefulness of the bar gene as a selectable gene for transformation. More recently, Christou *et al.* (1991) have regenerated transgenic rice plants using a CaMV 355 promoter-bar chimeric gene by electric discharge particle acceleration.

TRANSFORMATION SYSTEM

In dicotyledonous plant species, *Agrobacterium* mediated gene transfer is routine. Numerous examples we available in such species of successful gene transfer using *Agrobacterium* as a vector. In view of the absence of natural vectors in rice, transformation has been achieved mainly through protoplasts using direct DNA transfer methods. During the last 5 year, several reports have become available on regeneration of plants from protoplasts of both japonica and indica rice. In addition, over 10 cases, protoplasts have been successfully used to produce transgenic rice. The expression of foreign genes has been observed both at callus and plant levels. Since the first report in 1985 on plant regeneration from protoplasts of japonica rice, several laboratories have successfully regenerated plants from protoplasts of both japonica and indica rice. This has further let to the production of transgenic rice (Table l).

Table 1. Reported Instances of Foreign Gene Transfer in Rice Plants

Genotypes	Sources of protoplasts/ materials	DNA transfer methods	Genes transferred	References
Yamahoushi (J)	CS	Electroporation	NPT II	Toriyama *et al.* 1988
Taipei 309 (J), Pi4 (J)	CS	PEG	GUS	Zhang & Wu 1988
Taipei 309 (J)	CS	Electroporation	NPT II	Zhang *et al.* 1988
Fujisaka 5 (J)	Florets pathway	Pollen tube	NPT II	Luo & Wu 1989
Nipponbare (J)	CS	Electroporation	HPH	Shimamoto *et al.* 1989 Terada & Shimamoto 1989
Yamahoushi (J)	CS	Electroporation	HPH, GUS	Matsuki *et al.* 1989
Taipei 309 (J) Nipponbare (J)	CS	PEG	HPH	Hayashimoto 1990
Nipponbare (J) Taipei 309 (J)	CS	PEG	HPH	Li *et al.* 1990
IR 54 (I)	CS	PEG	NPT II	Peng *et al.* 1990
Chinsurah Boro II (I)	CS	PEG	HPH	Datta *et al.* 1990
Taipei 309 (J)	CS	Electroporation	NPT II, GUS	Battraw & Hall 1990
Nipponbare (J)	CS	Electroporation	HPH, GUS	Kyozuka *et al.* 1990

J; japonica, I; indica, CS; cell suspension, NPT II; neomycin phosphotransferase II, GUS; B-glucuronidase, HPH; hygromycin phosphotransferase

INHERITANCE OF FOREIGN GENES

The production of transgenic plants from rice protoplasts has been reported by several laboratories. With respect to the manner of foreign DNA integration, Shimamoto *et al.* (1989) presented evidence indicating Mendelian inheritance of foreign gene in the progenies of transgenic rice plants. The copy number of integrated *hph* gene was 2-10 per cell. Co- transformation of a non-selectable gene with a selectable marker was also recorded. Hayashimoto *et al.* (1990) regenerated transgenic rice plants, in which the introduced plasmid DNA could form concatemers by intermolecular recombination prior to integration. Datta *et al.* (1990) obtained transgenic plants of indica rice. They showed that 31 seeds derived from 5 primary transgenic plants were all hygromycin resistant, suggesting the primary transformants are homozygous. Using T_0 to T_3 generations we followed the fate of two genes which had been co-transferred into rice protoplasts (Goto *et al.* 1993). Our experiments confirmed Mendelian inheritance of the *hph* gene conferring hygromycin resistance. However, some plants did not show a clear segregation ratio. Such plants generally expressed weak resistance to hygromycin. Extensive analysis using several restriction enzymes indicates the random fragmentation of plasmid DNA. As expected, we were able to obtain progenies possessing hygromycin resistance at homozygous loci in T_2 and T_3 generations. Such plants also contained co-transferred counterpart DNA. This finding encourages utilisation of another gene co-transferred with a selectable hygromycin-resistance gene.

BIALAPHOS RESISTANT RICE PLANTS

We were able to detect PAT protein and PAT enzymatic activity in bialaphos resistant T_1 plants. These results verify transmission of the introduced Ubi-*bar* gene to the progeny of the primary transgenic rice plants. The availability of strong monocot promoters that are active in all tissues may aid the development of efficient transformation protocols that will routinely provide fertile, transgenic monocot plants at high frequency. The data presented here verify that the Ubi-1 promoter rives efficient expression of a selectable marker gene in rice allowing the regeneration of fertile, transgenic plants. When the leaf sheaths of two-months old transformed rice plants were inoculated with *R. solani* mycelia and examined two weeks later, plants that had been sprayed with an aqueous solution containing 200 mg/l of bialaphos (either three hours before or two days after inoculation) remained asymptomatic. Transgenic plants that were not treated showed typical symptoms of disease. In addition, penetration of fungal hyphae was also prevented in the bialaphos treatment plants. The dosage rate we used is approximately 10-times the minimal lethal dose to pathogens such as *R. solani* and *Pyricularia oryzae* and 10-times the minimal concentration needed to kill untransformed rice plants. Even two days after inoculation herbicide treatment was effective in suppressing disease symptoms, after three days, protection was only rarely observed. By a judicious choice of bialaplios concentration and time of application it may therefore be possible to combat both fungal infections and weed infestation simultaneously in fields of transgenic plants expressing a *bar* gene.

CONCLUSIONS

During the last 5 years, dramatic progress has been made in protoplast culture and DNA transformation of rice. Plant regeneration from protoplasts of both japonica and indica rices has been achieved in 10 number of laboratories. Various genes have been cloned in the form of cDNA as well as genomic DNA, comprehensive RFLP genetic map has become available and transposition of maize transposable element into rice has also recently been demonstrated. Plant regeneration from protoplasts is often limited to only specific genotypes and the results are often non-reproducible. Also, there is a need to have cloned genes governing useful agronomic traits such as disease and insect resistance, salinity and drought tolerance, and improved nutritional quality, etc.

Acknowledgments

This research was supported by grants from the Ministry of Education, Science, and Culture of Japan (H.U. & S.T.), the Ministry of Agriculture, Forestry and Fishery of Japan (H.U.), Suhara Memorial Foundation (H.U. & S.T.), Torey Science Foundation (H.U.), and the Rockefeller Foundation (H.U.).

REFERENCES

Battraw, M.J. and Hall, T.C. (1990). Histochemical analysis of CaMV 35S promoter-B-glucuronidase gene expression in transgenic rice plants. *Plant Mol Biol* **15**:527–538.

Bevan, M. Barnes, W.M. and Chilton, M. (1983). Structure and transcription of the nopaline synthase gene region of T-DNA. *Nucleic Acids Res* **1**: 369–385.

Christensen, A.H., Sharrock, R.A., and Quail, P.H. (1992). Maize polyubiquitin genes: structure, thermal perturbation of expression and transcript splicing, and promoter activity following transfer to protoplasts by electroporation. *Plant Mol Biol* **18**: 675–689.

Chu, C.C., Wang, C.C., Sun, C.S., Hsu, C., Yin, K.C., Chu, C.Y. and Bi, F.Y. (1975). Establishment of an efficient medium for anther culture of rice through comparative experiments on the nitrogen sources. *Sci Sin* **18**: 659–668.

Datta, S.K., Peterhans, A., Datta, K. and Potrykus, I. (1990). Genetically engineered fertile indica-rice recovered from protoplasts. *Bio/Technology* **8**:736–740.

De Block, M., De Brouwer, D. and Tenning P. (1989). Transformation of *Brassica napus* and *Brassica olercea* using *Agrobacterium tumefaciens* and the expression of the *Bar* and *neo* genes in the transgenic plants. *Plants Physiol* **91**:694–704.

De Block, M., Botterman, JU., Vandewiele, M., Dockyx, J., Thoen, C., Gossele, V., Movva, N.R., Thompson, C., Van Montagu, M. and Leemans, J. (1987). Engineering herbicide resistance in plants by expression of a detoxifying enzyme. *EMBO J* **6**:2513–2518.

Dekeyser, R., Claes, B., Marichal, M., Van Montagu, M., Caplan, A. (1989). Evaluation of selectable markers for rice transformation. *Plant Physiol* **90**:217–223.

Fromm, M.E., Morrish, F., Armstrong, C., Williams, R., Rhomas, J. and Klein, T.M. (1990). Inheritance and expression of chimeric genes in the progeny of transgenic maize plants. *Bio/Technology* **8**:833–839.

Gordon-Kamm, J.W., Spencer, T.M., Mangano, M.L., Adams, T.R., Daines, R.J., Start, W.G., O'Brien, J.V., Chambers, S.A., Adams Jr, W.R., Willetls, N.G., Rice, T.B., Mackey, C.J., Kruegar, R.W., Kausch, A.P. and Lemaux, P.G. (1990). Transformation of maize cells and regeneration of fertile transgenic plants. *Plant Cell* **2**:603–618.

Goto, G., Toki, S., and Uchimiya, H. (1993) Inheritance of a co-transferred foreign gene in the progenies of transgenic rice plants. *Transgenic Res.* **2**:300–305.

Hashida, S., Imagawa, M., Inoue, S., Ruan, K.H. and Ishikawa, E. (1984). More useful maleimide compounds for the conjugation of Fab′ to horseradish peroxidase through thiol groups in the hinge. *J. Appl. Biochem.* **6**:56–63.

Hayashimoto, A., Li, Z. and Murai, N. (1990). A PEG-mediated protoplast transformation system for production of fertile transgenic rice plants. *Plant Physiology* **93**:857–863.

Mastuki, R., Onodera, J., Yamauchi, T. and Uchimiya, H. (1989). Tissue-specific expression of the *rolC* promoter of Ri plasmid in transgenic rice plants. *Mol Gen Genet* **220**:12–16.

Murakami, T., Anzai, H., Imai, S., Satoh, A., Nagaoka, K. and Thomspson, C.J. (1986). The biolaphos biosynthetic genes of *Streptomyces hygroscopicus*: Molecular cloning and characterisation of the gene cluster. *Mol Gen Genet* **25**:42–50.

Murashige, T. and Skoog, F. (1962). A revised medium for rapid growth and bioassays with tobacco tissue cultures. *Physiol Plant* **15**:473–497.

Peng, J. Lyznik, L.A., Lee, L. and Hodges, T.K. (1990) Co-Transformation of indica rice protoplasts with *gus*A and *neo* genes. *Plant Cell Rep* **9**:168–172.

Potrykus, I. (1991) Gene transfer to plants: assessment of published approaches and results. Ann. Rev. Plant Physiol. *Plant Mol Biol.* **42**:205–225.

Shimamoto, K., Teda, R., Izawa, T. and Fujimoto, H. (1989). Fertile transgenic rice plants regenerated from transformed protoplasts. *Nature* **338**:274–277.

Shure, M., Wessler, S. and Fedoroff, N. (1983) Molecular identification and isolation of *waxy* locus in maize. *Cell* **35**:225–233.

Sugaya, S., Hawakawa, K., Handa, T. and Uchimiya, H. (1989) Cell specific expression of the *rolC* gene of the TL-DNA of Ri plasmid in transgenic tobacco plants. *Plant Cell Physiol* **30**:649–653.

Tachibana, K., Watanabe, T., Sekizawa, Y. and Takematsu, T. (1986a). Inhibition of glutamine synthetase and quantitative changes of free amino acids in shoots of bialaphos-treated Japanese barnyard millet. *J Pesticide Sci* **11**:27–31.

Tachibana, K., Watanabe, T., Sekizawa, Y. and Takematsu, T. (1986b) Accumulation of ammonia in plants treated with bialaphos. *J. Pesticide Sci* **11**:33–37.

Tada, Y., Sakamoto, M. and Fujimur, T. (1990). Efficient gene introduction into rice by electroporation and analysis of transgenic plants: use of electroporation buffer lacking chloride ions. *Theor Appl Genet* **80**:475–480.

Tada, Y., Sakamoto, M., Matsuoka, M. and Fujimura, T. (1991). Expression of a monocot LHCP promoter in transgenic rice. *EMBO J* **7**:1803–1808.

Terada, R. and Shimamoto, K. (1990). Expression of CAMV35S-GUS gene in transgenic rice plants. *Mol Gen Genet* **220**:389–392.

Thompson, C.J., Movva, N.R., Tizard, R., Crameri, R., Davies, J.E., Lauwereys, M. and Botterman, J. (1987). Characterisation of the herbicide-resistance gene *bar* from *Streptomyces hygroscopicus*. *EMBO J* **6**:2519–2523.

Toriyama, K., Hirata, K. (1985). Cell suspension and protoplast culture in rice. *Plant Sci* **41**:179–183.

Toriyama, K., Arimoto, Y., Uchimiya, H. and Hinata, K. (1988). Transgenic Rice Plants after Direct Gene Transfer into Protoplasts. *Bio/Technology* **6**:1072–1074.

Zhang, H.M., Yang, H., Rech, E.L., Golds, T.J., Davis, A.S., Mulligan, B.J., Cocking E.C. and Davey, M.R. (1988). Transgenic rice plants produced by electroporation-mediated plasmid uptake into protoplasts. *Plant Cell Rep* **7**:379–384.

GENETIC ENGINEERING OF OAT

D. A. Somers, K. A. Torbert, W. P. Pawlowski, and H. W. Rines

Department of Agronomy and Plant Genetics
University of Minnesota, St. Paul, Minnesota 55108 and

Plant Science Research Unit
U.S. Department of Agriculture,
Agriculture Research Service
St. Paul Minnesota 55108

SUMMARY

Fertile, transgenic oat (*Avena sativa* L.) plants were regenerated from approximately 35% of phosphinothricin (PPT)-resistant callus cultures selected following microprojectile bombardment to deliver the plasmid pBARGUS. The plasmid pBARGUS contains the *bar* gene, which confers plant cell resistance to PPT and related herbicides, and the *uidA* gene for β-glucuronidase (GUS) under the control of the maize alcohol dehydrogenase I promoter. This promoter conferred high levels of GUS activity in the endosperm of mature oat kernels, thus enabling ready determination of transmission genetics of the GUS transgene. Segregation ratios of GUS activity in mature R_1 seed and, in some cases, the R_2 and R_3 generations were determined in 15 transgenic families. Seven families fit a 3:1 GUS^+:GUS^- segregation ratio whereas two families segregated 15:1 for GUS activity. The remaining six families exhibited aberrant segregation ratios. These initial studies used friable, embryogenic callus initiated from immature embryos of a specific genotype selected for high frequency callus initiation. Although this system was useful in establishing and characterizing oat transformation, its limitation to a specific genotype and the undesirability of herbicide tolerance in oat dictated further development of transformation systems for use in oat improvement. Transformation of current oat cultivars and the use of an antibiotic-based selection system to obviate the herbicide resistance marker are described.

INTRODUCTION

Production of fertile, transgenic oat (*Avena sativa* L.) plants was first reported in 1992 (Somers et al. 1992). In that report, friable embryogenic (FE) callus and embryogenic

Improvement of Cereal Quality by Genetic Engineering, Edited by
Robert J. Henry and John A. Ronalds, Plenum Press, New York, 1994

suspension cultures initiated from a highly culturable genotype were bombarded with microprojectiles coated with the plasmid pBARGUS. In the three experiments described, only 33 of 111 transgenic callus lines exhibited plant regeneration capacity and the majority of regenerated plants were either completely sterile or male sterile. Plants regenerated from two transgenic callus lines were completely fertile. Transgene sequences and phenotypes (seedling PPT-resistance and β-glucuronidase (GUS) activity in mature seed) were transmitted to the progeny of regenerated plants and cosegregated at ratios indicative of disomic inheritance of the transgenes (Somers et al. 1992). Suspension cultures and FE callus produced similar frequencies of transgenic plants per microprojectile bombardment treatment. Because suspension cultures appeared to offer no advantages over callus as a target tissue, and exhibited the disadvantages of prolonged development times and ensuing infertility of regenerated plants, FE callus was subsequently used as a source of totipotent target cells. The low recovery of fertile, transgenic oat plants in these initial experiments warranted further research to improve the transformation system. Accordingly, we recently have concentrated on bombarding younger FE oat callus. The results of this effort coupled with improvement in selection procedures have yielded a large increase in recovery of fertile, transgenic oat plants. The objectives of this report are to (1) describe the current status of oat transformation, (2) report on transgene inheritance data, and (3) discuss our current efforts to improve the transformation system for applications to oat improvement.

MATERIALS AND METHODS

Oat Tissue Culture

Friable, embryogenic (FE) callus was initiated from immature embryos (Bregitzer et al. 1989). The majority of our work employs a genotype referred to as GAF that was selected for elevated levels of callus initiation. GAF is an F4 derivative of a cross between the *Avena sativa* cultivar 'Garland' and the wild oat (*Avena fatua*) accession 1223. Other genotypes that are currently being investigated are 'Starter' and 'Donald.' Immature embryos were cultured on either a callus initiation medium referred to as MS-2D consisting of MS salts (Murashige and Skoog 1962) containing 2 mg/L 2,4-dichlorophenoxyacetic acid (2,4-D) (Bregitzer et al. 1989) or a more recently developed MS-based medium containing 0.2 mg/L 2,4-D and 0.2 mg/L benzylaminopurine (BAP) (Milach et al. 1992). Following initiation on either medium, callus was maintained on MS-2D and was continually selected for FE appearance and for rapid, stable growth. This selection process takes up to 3 months. For plant regeneration, callus was transferred to regeneration medium consisting of MS salts containing 2 mg/L napthyleneacetic acid and 0.2 mg/L BAP to promote shoot formation followed by transferring shoots to MS medium containing no growth regulators for root formation (Bregitzer et al. 1989).

Delivery of Plasmid DNA and Selection of Transgenic Tissue Cultures

Plasmid DNA was delivered using the gunpowder Biolistic PDS-1000 microprojectile bombardment device (Somers et al. 1992). FE oat callus was the target tissue. Plasmid DNA was coated onto tungsten M-17 microprojectiles according to Finer et al. (1992).

Two types of plasmids were utilized for selection of transgenic oat tissue cultures. The plasmid pBARGUS (Fromm et al. 1990) carries the *Streptomyces hygroscopicus bar* selectable marker gene (Thompson et al. 1987) under the control of the cauliflower mosaic virus (CaMV 35S) promoter combined with the maize alcohol dehydrogenase(Adh) intron I enhancer sequence, and the β-glucuronidase (GUS) reporter gene (Jefferson et al. 1987) under the control of the Adh promoter-intron I enhancer. Selection of transgenic tissue cultures that were treated with pBARGUS was conducted using MS-2D medium lacking asparagine and containing either 3 mg/L phosphinothricin (Central Chemical Co. Hauppauge, NY, USA) (Somers et al. 1992) or bialaphos (generously provided by Meiji Sieko Kiasha, Ltd., Yokohama, Japan). The plasmid pH24 (generously provided by Dr. M.E. Fromm, Monsanto Co., St. Louis, MO, USA) carries the neomycin phosphotransferase (NPT II) gene as the selectable marker under control of the CaMV 35S promoter-AdhI Intron 1 enhancer cassette. Tissue culture selection using the NPT II selectable marker gene constructs was conducted using MS-2D medium containing from 50 to 150 mg/L paromomycin. In cotransformation experiments, equal concentrations of two plasmids were combined prior to DNA delivery.

Analysis of Transgenic Tissue Cultures

Following approximately 3 months of tissue culture selection, surviving tissue cultures were transferred to plant regeneration medium containing the appropriate selective agents. At this stage of the experiments, callus was stained for GUS activity (Kosugi et al. 1990) if the tissues had been bombarded with pBARGUS. The tissue cultures bombarded with pH24 and selected on medium containing paromomycin were assayed for the NPT II protein using a NPT II ELISA assay kit (5 prime → 3 prime, Inc., Boulder, CO, USA). All selected callus lines and leaf tissue from some regenerated plants were freeze dried and DNA was extracted (Saghai-Maroof et al. 1988). Southern blot analyses (Southern 1975) were conducted on unrestricted and restricted oat DNA. Membranes were hybridized with *bar* or *npt II* gene probes as described by Somers et al. (1992).

Transgene Inheritance

Multiple plants were regenerated from the selected tissue cultures. Plants regenerated from the same transgenic callus were referred to as a family. The Adh I promoter, which controls expression of the GUS gene in the introduced plasmids, confers high levels of GUS activity in mature cereal grains (Kyozuka et al. 1991). Oat groats were cut in half and the endosperm portion was stained for GUS activity (Kosugi et al. 1990). The embryo portion was planted for further analyses. Expression and segregation of GUS activity were determined in self-pollinated progeny of regenerated plants. In some cases, seedling leaves were treated with 0.2% active ingredient PPT by applying the herbicide IGNITE to determine herbicide tolerance phenotype (Gordon-Kamm et al. 1990). R_1 and R_2 plants were bagged to control outcrossing.

RESULTS AND DISCUSSION

Transformation Experiments

During 1992, six transformation experiments were conducted from which 116 PPT-resistant tissue cultures were selected. PPT-resistant callus lines selected in four of these experiments have been analysed using Southern blots for detection of the *bar* gene. Plants were regenerated from these transgenic callus lines and characterized (Table 1),. thus enabling a current assessment of the oat transformation protocol. GAF callus less than one-

Table 1. Summary of 4 oat transformation experiments.

Experiment Fertile Number Plants	Southern Positive	Number of Tissue Cultures GUS Activity	Cotransformed*	Plant Regeneration
500	6	3	5/6	64
600	8	7	3/4	74
800	15	11	4/10	84
900	4	4	2/4	3
Totals	33	25 (76%)†	14/24 (58%)	24 (73%)

* In Experiment #500, callus was treated with pBARGUS and pH24; NPT II protein was detected by ELISA. In Experiments 600, 800 and 900, callus was treated with pBARGUS and barley yellow dwarf virus coat protein (BYDV-CP) constructs; BYDV-CP genes were detected by polymerase chain reaction. Data are numbers of cotransformed tissue cultures per tissue cultures analyzed.

† Numbers in parenthesis are overall percentages of transgenic callus in each category.

year-old was bombarded. All experiments utilized the pBARGUS plasmid with selection on either PPT or the related compound bialaphos. Furthermore, all experiments were cotransformation experiments. The first experiment (#500) involved cointroduction of pBARGUS and pH24; the remaining three experiments cointroduced pBARGUS and barley yellow dwarf virus coat protein (BYVD-CP) constructs. This project is being conducted in collaboration with R. Lister, Purdue University, to investigate resistance mechanisms to BYDV. R1 and R2 transgenic plants are currently being analysed for resistance to BYDV. In this presentation, these four experiments will be reported to characterize the cotransformation frequency of the oat transformation system.

The results of the four completed experiments are shown in Table 1. A total of 37 PPT-resistant tissue cultures selected in these experiments have been analysed for the presence of the *bar* gene and 33 callus lines were transgenic indicating that 89% of the PPT-resistant tissue cultures were transgenic. Of the transgenic tissue cultures, 76% expressed GUS activity. Over all four experiments, 24 of the transgenic tissue cultures regenerated plants and of these, 13 tissue cultures regenerated at least one completely fertile plant. Over all experiments, transgenic tissue cultures that regenerated fertile, transgenic plants were produced at a frequency of 0.15 per microprojectile bombardment.

In Experiment #500, 5 out of 6 of the transgenic callus lines were cotransformed with pH24 as determined by detection of the NPT II protein using ELISA (Table 1). Cotransformation of the BYDV-CP genes in Experiments 600, 800 and 900 was determined by polymerase chain reaction. In the three BYDV-CP experiments, 9 of the 18 transgenic tissue cultures analyzed exhibited BYDV-CP sequences. Of the 9 tissue cultures cotransformed with BYDV-CP, four have regenerated plants that were self-fertile (data not shown). The high frequency of cotransformation (58% overall) indicates that microprojectile bombardment will be useful for transforming oat with genes of agronomic interest by cointroduction with the selectable marker genes.

Transgenes were integrated into the oat genome based on comparisons of Southern analyses of restricted and unrestricted DNA isolated from the PPT-resistant tissue cultures using the *bar* gene as a probe. Copy number was estimated to range from one to more than

20 copies of the *bar* gene per haploid genome in the transgenic tissue cultures (data not shown). Transgene rearrangements as indicated by multiple hybridizing fragments both larger and smaller than the unit length of the *bar* gene were frequently observed. In our previous studies of the transgenic tissue cultures that did not express GUS activity, generally no unit length GUS (*uidA*) gene was detected indicating that some aspect of transgene integration led to deletion or at least mutation of these linked transgene sequences. We are in the process of conducting a detailed analysis of transgene integration using the *bar*, *uidA* and cotransformed genes as probes. It is likely that transgene rearrangements may result from fragmentation during DNA delivery, and concatamerization and breakage during integration. We also are in the process of examining the transgene integration patterns in multiple plants regenerated from each of the transgenic tissue cultures to investigate the degree of chimerism of transgene integration within the selected tissue cultures and to assess the stability of transgene integration patterns during the plant regeneration process. It would be of interest to determine if any relationship exists between the transgene integration pattern (copy number and rearrangements) and the transmission genetics of the transgene(s). Furthermore, the effect of transgene integration pattern on gene expression levels will be investigated. We believe that further analyses of the transgenic tissue cultures isolated in these experiments will provide useful information for strategies to maximize recovery of transgenic plants that transmit transgenes to their progeny in normal Mendelian fashion.

Transgene Inheritance

In our initial studies characterizing progeny of transgenic plants (Somers et al. 1992), we observed that transgene sequences cosegregated with the transgene phenotypes (GUS activity in seed and PPT-resistance in seedlings). We are in the process of conducting a similar analysis on the fertile families from the 1992 experiments and two transgenic families produced from previous experiments.

Regenerated plants (Ro) derived from transgenic tissue cultures isolated in each experiment were allowed to self-pollinate. Transgene segregation was evaluated in progeny of plants regenerated from 15 GUS-positive (GUS^+), PPT-resistant, transgenic tissue cultures. R_1 progeny (seed produced on regenerated plants) and, in some cases, their corresponding self-pollinated R_2 and R_3 families were evaluated by determining GUS expression in mature groats. In some cases PPT-resistance in seedlings also was determined. Transgenic tissue cultures were designated according to experiment, e.g., 500, 502, etc. from Experiment #500, and grouped in Tables 2abc based on segregation ratios for GUS activity. Seven of the 15 R_1 families exhibited close fit to a 3:1 GUS^+:GUS^- segregation ratio (Table 2a) indicating disomic inheritance of a single dominant locus for GUS activity. Two R_1 families segregated 15:1 for GUS activity suggesting two independent GUS loci (Table 2b); whereas, the remaining 6 families exhibited segregation of GUS activity that deviated significantly from these ratios (Table 2c). R_2 seed produced on self-pollinated R_1 plants regenerated from 7 transgenic tissue cultures were assayed for GUS activity. In this experiment (Table 2abc), at least 8 R_1 plants were produced from each family to maximize recovery of homozygous transgenic R_2 families. R_1 genotype was determined based on GUS assays of at least 20 R_2 seeds per plant in most cases. Homozygous GUS^+ R_2 families were recovered in 4 of the 7 families tested indicating stable transgene inheritance. The total R_1 transgene segregation ratios for lines 601 and 804 based on R_2 progeny tests conformed to an expected 1:2:1 ratio (Table 2a). The R_1 GUS segregation ratio of each family was generally predictive of the transgene genotype (1 or 2 loci, or aberrant segregation) determined by the R_2 progeny tests.

Table 2a. Segregation of GUS activity in self-pollinated progeny (seed) of transgenic oat plants exhibiting close fit to a 3:1 GUS^+:GUS^- ratio. All segregating families conformed to a 3:1 segregation ratio at $P < 0.05$.

Family	Generation	Number of Plants	Number of Progeny	
			GUS^+	GUS^-X^2
300	R_1*	1	233	830.27
	R_2 GUS^+/GUS^+	6	295	0
	R_2 GUS^+/GUS^-	13	377	1280.03
504	R_1	1	15	50
506	R_1	1	19	50.22
601	R_1	5	173	660.87
	R_2 GUS^+/GUS^+	4	63	0
	R_2 GUS^+/GUS^-	16	205	690.005
	R_2 GUS^-/GUS^-	9	2	160
803	R_1	1	12	82.40
804	R_1	2	101	380.41
	R_2 GUS^+/GUS^+	5	70	0
	R_2 GUS^+/GUS^-	15	199	812.30
	R_2 GUS^-/GUS^-	7	1	117
	R_1†	1	0	50
809	R_1	2	28	120.533

* Only GUS^+ R_1 seed were planted in this family.

† This GUS^- R_1 plant suggests that the 804 callus was chimeric for the presence of the transgene.

Table 2b. Segregation of GUS activity in self-pollinated progeny (seed) of transgenic oat plants exhibiting close fit to 15:1 GUS^+:GUS^- ratio

Family	Generation	Number of Plants	Number of Progeny	
			GUS^+	GUS^-X^2
900	R_1	2	36	40.96
	R_2 GUS^+/GUS^+	4	80	0
	R_2 GUS^+/GUS^-	5	68	9
	R_2 GUS^-/GUS^-	0		
903	R_1	1	29	10.44

Table 2c. Aberrant segregation from GUS activity in progeny (seed) of transgenic oat plants

Family for 3:1	Generation	Number of Progeny		
		Number of Plants	GUS^+	GUS^-
400 (5.88*)	R_1 †		64	36
	R_2 GUS^+/GUS^+	0		
	R_2 GUS^+/GUS^-	7	243	255
	R_3	12	153	220
	R_1	1	95	26
	R_2 GUS^+/GUS^+	0		
	R_2 GUS^+/GUS^-	24	206	169
	R_1	1	54	24
	R_1	1	51	23
500 (40.42*)	R_1	2	33	43
	R_2 GUS^+/GUS^+	0		
	R_2 GUS^+/GUS^-	12	111	108
	R_2 GUS^-/GUS^-	10	0	120
503 (13.44*)	R_1	2	25	23
604 (6.667*)	R_1	1	10	10
607 (5.40*)	R_1	1	10	10
610	R_1	1	14	2
	R_2 GUS^+/GUS^+	0		
	R_2 GUS^+/GUS^-	1	2	1
	R_2 GUS^-/GUS^-	2	0	23

* Indicates significantly different from a 3:1 segregation ratio at $P < 0.05$.
† Only GUS^+ R_1 seed were planted in this family.

The aberrant segregation ratios shown in Table 2c are unavoidably more interesting than normal transgene segregation. R_1 segregation in these six lines either appeared to approximate 3:1 or tended more towards 1:1 segregation for GUS^+:GUS^-. R_2 segregation ratios of three of these lines were closer to 1:1 segregation for GUS and no transgenic homozygotes were recovered in the R_2 or the R_3 generations of the families tested. We speculate that in these cases the transgene may not be transmitted through the male gametes. If this hypothesis is true, it may indicate association of the transgene with a chromosomal aberration. Cytological and molecular analyses will be required to understand the mechanisms underlying these apparently unstable events.

Cosegregation of PPT-resistance and GUS activity which was expected due to the adjacent linkage of the *bar* and *uidA* genes on the pBARGUS plasmid was tested in R_1 progeny regenerated from five transgenic tissue cultures. Seedling PPT-resistance appeared to cosegregate with seed GUS activity in 3 of the 5 tested families (Table 3). The few apparent

Table 3. GUS activity and PPT resistance in transgenic oat plants. GUS assays were conducted on excised endosperm of mature groats. PPT-resistance was determined on seedlings grown from GUS stained seed

Family	Regenerated Plant	GUS^+		GUS^-	
		PPT^+	PPT^-	PPT^+	PPT^-
601	1	44	2	2	17
903	1	13	1	0	0
900	2	11	1	1	0
503	1	11	0	15	6
804	2	16	21	1	19

recombinants in families 601, 900 and 903 were most likely due to problems with determining PPT-resistance using the swab method. In 2 families (503 and 804), apparent recombination between PPT and GUS was observed; e.g., GUS^+ seed that produced PPT-sensitive seedlings and GUS^- seed that gave rise to PPT-resistant seedings. In both families, deviations from Mendelian segregation of at least one of the transgene phenotypes were observed. These events could be due to multiple insertion loci for the transgenes followed by uncharacterized events that cause either deletion, mutation or other mechanisms that silenced one or both of the transgene loci. We plan to conduct further cosegregation tests of GUS activity and PPT-resistance in all fertile families to characterize the overall frequency of these events. Based on our uncertainty about the PPT-resistance scoring, we will conduct these analyses in R_2 progeny to determine R_1 genotype. Molecular analysis of these events will be conducted to investigate the cause of the aberrant segregation ratios. It will also be of interest to eventually investigate the cosegregation of PPT-resistance, GUS activity, and the genes (either NPT II or BYDV-CP) cotransformed from the separate plasmids.

Advantages and Disadvantages of the Oat Transformation Protocol

A reproducible oat transformation system is now available for applications of transformation to oat improvement. Overall, about one transgenic tissue culture that produced fertile plants was isolated per seven microprojectile bombardments of FE callus using this procedure. About 89% of the PPT-resistant tissue cultures analysed to date were transgenic based on detection of the *bar* gene indicating that selection using PPT is efficient. Integration of cotransferred genes on a plasmid separate from the selectable marker genes was detected in 58% of the transgenic tissue cultures. Bombardment as early as 3 months following callus initiation has yielded transgenic callus. Furthermore, transgenic tissue cultures have been isolated following microprojectile bombardment of several independently initiated callus cultures of the GAF genotype. These features enable us to pursue oat transformation studies throughout the year with high probability of recovering fertile, transgenic plants. Current applications include investigations of resistance to BYDV using BYDV-CP constructs and tissue specificity of Badnavirus (commelina yellow mottle) promoter sequences. Longer term projects are focused on resistance mechanisms to fungal pathogens and eventually on grain quality.

Transgenic herbicide resistance and use of a highly culturable oat genotype are integral features that pose the greatest disadvantages for the oat transformation system described in this chapter. The most serious drawback arises from the use of the *bar* gene as a selectable marker with the end result that the transgenic plants are resistant to bialaphos-and PPT-containing herbicides. Oat is sexually compatible with its weedy relatives, *Avena sterilis* and *A. fatua*. Outcrossing of transgenic oat with wild oat is a definite eventuality and, if the transgenic cultivated oats are herbicide resistant, this trait will be transferred to wild oat. Herbicide-tolerant wild oat would pose a serious weed problem in crops such as wheat that have been genetically engineered to be resistant to PPT-containing herbicides (Vasil et al. 1992). In attempting to develop alternative systems for selection of transgenic oat, we have found that paromomycin, a neomycin derivative, in combination with NPT II introduced on the plasmid pH24 provided efficient selection of transgenic oat tissue cultures (Torbert and Somers, unpublished results). An average of approximately 3 paromomycin-resistant tissue cultures were isolated per microprojectile bombardment in two separate experiments. The tissue cultures and their regenerated plants were shown to be transformed based on detection of the NPT II protein using an ELISA assay, and by detection of the NPT II structural gene in Southern blot analyses (data not shown). Regenerated plants are nearing anthesis and will be evaluated for fertility. We anticipate good fertility because the tissue cultures used for these experiments were of similar age and genotype as in our previous experiments using PPT selection. We believe that NPT II transgenic plants will pose reduced risk in field tests compared with PPT-resistant plants. However, we recognize the concerns expressed about using antibiotic resistance as a selectable marker. Accordingly, our longer term goals are to investigate strategies to eliminate the selectable marker gene from transgenic oat plants.

The usefulness of this transformation system in oat improvement depends on its applicability to oat cultivars. Progress recently has been achieved in initiating regenerable callus from oat cultivars (Milach et al. 1992). In preliminary experiments, tissue cultures initiated from the cultivars Starter and Donald using procedures developed by Milach et al. (1992) were bombarded with the plasmid pH24 and selected on paromomycin-containing medium. Southern analyses and NPT II ELISA on paromomycin-resistant Donald tissue cultures indicate that they are transgenic. This result is very promising because it indicates that manipulation of callus initiation medium will result in establishment of transformable callus in oat cultivars. Further analyses will include determining fertility of plants regenerated from the transgenic Donald tissue cultures and transgene inheritance.

Acknowledgements

The authors wish to express their gratitude to Dr. M.E. Fromm (Monsanto Co., St. Louis, MO) for providing the plasmids used in this research. This project was supported in part by grants from The Quaker Oats Co. and the Midwest Plant Biotechnology Consortium, USDA subgrant 3593-0009-04.

Cooperative investigation of the Minnesota Agricultural Experiment Station and the USDA-Agricultural Research Service.

Mention of a trademark, vendor, or proprietary product does not constitute a guarantee or warranty of the product by the United States Department of Agriculture or the University of Minnesota, and does not imply its approval to the exclusion of other products or vendors which may also be suitable.

REFERENCES

Bregitzer, P., Somers, D. A. and Rines, H. W. (1989) Development and characterization of friable, embryogenic oat callus. *Crop Sci.* 29, 798–803.

Finer, J. J., Vain, P., Jones, M. W. and McMullen, M. D. (1992) Development of the particle inflow gun for DNA delivery to plant cells. *Plant Cell Rep.* 11, 323–328.

Fromm, M. E., Morrish, F., Armstrong, C., Williams, R., Thomas, J. and Klein, T. M. (1990) Inheritance and expression of chimeric genes in the progeny of transgenic maize plants. *Bio/Technology* 8, 833–839.

Gordon-Kamm, W. J., Spencer, T. M., Mangano, M. L., Adams, T. R., Daines, R. J., Start, W. G., O'Brien, J. V., Chambers, S. A., Adams, W. R., Willetts, N. G., Rice, T. B., Mackey, C. J., Krueger, R. W., Kausch, A. P. and Lemaux, P. G. (1990) Transformation of maize cells and regeneration of fertile transgenic plants. *Plant Cell* 2 , 603–618.

Jefferson, R.A. 1987. Assaying chimeric genes in plants: the GUS gene fusion system. *Plant Molec. Biol. Rep.* 5, 387–405.

Kosugi, S., Oshashi, Y., Nakajima, K. and Arai, H. (1990) An improved assay for β-glucuronidase in transformed cells: methanol almost completely suppressed endogenous β-glucuronidase activity. *Plant Sci.* 70, 133–140.

Kyozuka, J., Fujimoto, H., Izawa, T. and Shimamoto, K. (1991) Anaerobic induction and tissue specific expression of maize *Adh1* promoter in transgenic rice plants and their progeny. *Mol. Gen. Genet.* 228, 40–48.

Milach, S. C. K., Rines, H. W., Somers, D. A., Gu, W. and Grando, M. (1992) Improvements in embryogenic callus cutlure in oat (*Avena sativa* L.) In: *International Crop Science Congress Abstracts.* Iowa State University, Ames, IA. p. 61.

Murashige, T. and Skoog, F. (1962) A revised medium for rapid growth and bioassays with tobacco tissue cultures. *Physiol. Plant* 15, 473–397.

Saghai-Maroof, M. A., Soliman, K. M., Jorgensen, R. A., and Allard, R. W. (1984) Ribosomal DNA spacer-length polymorphisms in barley: Mendelian inheritance, chromosomal location, and population dynamics. *Proc. Natl. Acad. Sci. (USA)* 81, 8014–8018.

Somers, D. A., Rines, H. W., Gu, W., Kaeppler, H. F. and Bushnell, W. R. (1992) Fertile, transgenic oat plants. *Bio/Technology* 10, 1589–1594.

Southern, E. M. (1975) Detection of specific sequences among DNA fragments separated by gel electrophoresis. *J. Mol. Biol.* 98, 503–517.

Thompson, C. J., Movva, N. R., Tizard, R., Crameri, R., Davies, J. E., Lauwereys, M. and Botterman, J. (1987) Characterization of the herbicide-resistance gene *bar* from *Streptomyces hygroscopicus. EMBO J.* 6, 2519–2523.

Vasil, V., Castillo, A. M., Fromm, M. E. and Vasil, I. K. (1992) Herbicide resistant fertile transgenic wheat plants obtained by microprojectile bombardment of regenerable, embryogenic callus. *Bio/Technology* 10, 667–674

TRANSGENIC GRAIN SORGHUM (*Sorghum bicolor*) PLANTS VIA *Agrobacterium*

Ian Godwin and Rachel Chikwamba*

Department of Agriculture
The University of Queensland
Brisbane, Queensland 4072
Australia

SUMMARY

The ability to transfer foreign genes to grain sorghum would potentially allow production of cultivars with improved insect resistance and grain quality (nutritional, baking and brewing). We have established an *Agrobacterium*-mediated transformation protocol, based predominantly on axenic seedlings of inbred lines. After inoculation of *Agrobacterium tumefaciens* (LBA4404 pBI121 and AGLO pKIWI105) into wounded coleoptiles of *in vitro* germinating seedlings, we have generated putative transformants. No selection for antibiotic resistance was imposed. Expression of the GUS marker gene has been detected histochemically in 23 plantlets (from 250 inoculated seedlings). Histochemical GUS expression in non-transformed control sorghum seedlings has not been observed. Two individuals express GUS in all plant parts tested, while some of the remainder have whole tillers expressing the marker gene. Five primary transformants and their selfed progeny have been examined for GUS expression. In most cases, segregation ratios depart from the expected 3:1 ratio, possibly due to chimaerism or gene methylation. Southern analysis has revealed hybridisation to GUS sequences in high molecular weight genomic DNA in both primary transformants and selfed progenies.

INTRODUCTION

Grain sorghum (*Sorghum bicolor* L. Moench.) is grown throughout the tropics and sub-tropics as a stockfeed, and in drier parts of Africa and Asia is a staple food, with other important uses including brewing and syrup production (Purseglove 1972). Some of the lim-

---

*Present address: ENDA-Zimbabwe, P.O. Box 3492, Harare, Zimbabwe itations to grain yield and quality could be addressed via a genetic engineering approach. Traits such as pest and disease resistance, and altered grain protein quality would be achievable targets if a routine genetic transformation system was available for sorghum. Transgenic rice (Shimamoto *et al.* 1989), maize (Rhodes *et al.* 1988; Fromm *et al.* 1990), wheat (Vasil et al. 1992) and oats (Somers *et al.* 1992) have been produced by direct gene transfer protocols including electroporation of protoplasts and micro-projectile bombardment. To date, there is no fully substantiated report of transgenic sorghum.

For most dicotyledonous species, *Agrobacterium*-mediated gene transfer remains the preferred method. This is due in part to the simplicity and efficiency of the system, utilizing the natural ability of this soil-borne pathogenic bacterium to transfer segments of DNA to plant cells, where this DNA is then stably incorporated into the nuclear genome. The *Agrobacterium*-mediated approach is also generally the most economical gene transfer available. Until recently however, it had been widely assumed that the economically important cereals were beyond the host-range of *Agrobacterium tumefaciens*, and that direct gene transfer methods would be required to genetically engineer cereals. It now appears that *Agrobacterium*-mediated transformation can be achieved in cereals, with reports of transgenic rice tissues (Raineri *et al.* 1990) and plants (Chan *et al.* 1993) and maize plants (Gould *et al.* 1991) via the *Agrobacterium* pathway. Transformation of sorghum tissues, via electroporation of protoplasts (Battraw and Hall 1991) and micro-projectiles (Hagio *et al.* 1991) has been achieved, and a brief report of *Agrobacterium*-mediated transformation exists (Smith, 1991).

Whether *Agrobacterium*-mediated or direct gene transfer methods are used, a means of regenerating plants from cell or tissue cultures is generally necessary. Such techniques are available for sorghum, with plant regeneration possible from a range of explant tissues including seedling nodes (Masteller and Holden 1970) immature embryos (Dunstan *et al.* 1979), immature inflorescences (Brettel *et al.* 1980) and leaf bases (Wernicke and Brettel 1980). Regeneration is also possible from shoot apex cultures (Bhaskaran and Smith 1988). In all cases however, regeneration of sufficient frequency for efficient gene transfer is limited to a small number of genotypes.

We present here, a simple, rapid method for producing sorghum primary tranformants, which does not rely on the ability to regenerate from cell and tissue culture.

MATERIALS AND METHODS

Plant Material

Seeds of the inbred lines TAM422, QL27, Ajebsido and Tx430 were surface sterilised for 15 minutes in commercial bleach (Domestos, 5% sodium hypochlorite), rinsed three times in sterile distilled water and placed on MS medium (2% sucrose, 0.8% agar at pH 5.8) in petri dishes. These were placed in a dark cupboard 27°C and allowed to germinate. Seedlings were inoculated just prior to coleoptile emergence from the pericarp (variable 3–5 days).

Bacterial Cultures

The *Agrobacterium tumefaciens* strains used were LBA4404 with binary plasmidpBI121 (kindly donated by R. Jefferson, CSIRO Plant Industry, Canberra) and Agl0 with binary plasmid pKIWI105 (kindly donated by R. Gardner, Auckland University). Inocula were grown in LB broth containing 25 mg l^{-1} kanamycin. Cultures were incubated for 24 h at 27°C, 250 r.p.m. until they contained 10^9 cells ml^{-1} as determined by optical density.

Seedling Inoculation

Seedling inoculation was carried out in a laminar flow cabinet under a dissecting microscope. A hypodermic needle (24G) dipped in bacterial suspension was used to introduce the bacteria into the emerging coleoptile. The inoculated seedlings were placed on MS medium (2% sucrose, 0.8% agar at pH 5.8) and co-cultivation allowed for two days, after which they were transferred to the same MS medium supplemented with 500mg l^{-1} cefotaxime to prevent proliferation of the bacteria. No selective agent was used because seedlings were found to grow on 500 mg l^{-1} kanamycin. After seven days on antibiotic the seedlings were transferred to pots in steam sterilised soil in the glasshouse. Plants were grown to maturity with inflorescences bagged to ensure self-pollination.

Histochemical GUS Assays

The histochemical assays were carried out on 20 day old plants to detect the activity of the β-glucuronidase enzyme using the procedure according to Jefferson (1987). Pieces of leaf tissue devoid of necrotic spots were sectioned thinly, placed in glass tube and incubated at room temperature in a few drops of X-Gluc (5-bromo-4-chloro-3-indolyl glucuronide). After 4–5 hours, the leaf tissues were examined microscopically for the blue precipitate produced in the presence of the GUS enzyme. In some cases, the reaction was allowed to proceed overnight for positive identification of GUS activity.

Progeny Tests

Seeds from each tiller were treated as separate populations to account for chimaerism. A sample of 30 seeds (or in some cases as low as 9 due to low seed set) was surface sterilised (15 minutes in neat Domestos, with 3 rinses in sterile distilled water) and germinated in petri dishes on sterile filter paper. The 4–5 day old seedlings were then transferred into steam sterilised soil and grown in a controlled environment cabinet at 30°/22°C (day/night) with a 14 hour day. Leaf pieces were collected from 7–20 day old seedlings and histochemical assays were carried out according to Jefferson (1987). The leaf sections were placed in a vacuum dessicator for 3–5 minutes to encourage infiltration of the substrate. After 4–5 hours, leaf sections were examined for the blue precipitate.

DNA Analysis

DNA extraction was performed from young leaves using a mini-prep method modified after Dellaporta *et al.* (1983). Genomic DNA was restricted with *Hind*III which cuts once within the T-DNA, and electrophoresed in agarose (0.7%) gel. DNA alkali blot analysis was performed as described in Sambrook *et al.* (1989) on nylon membrane (Hybond N+, Amersham). The GUS probe was isolated from pBI101 (R.A. Jefferson) as a *Hind*III-*Eco*RI fragment, and radiolabelled with [α^{32}P]-dCTP using the random primer method (Feinberg and Vogelstein, 1983).

RESULTS AND DISCUSSION

The post-inoculation survival rate of seedlings varied in the range of 50–60% (Figure 1). This was attributed to the damage done to the meristem during inoculation. In most cases, seedling death was associated with production of phenolic compounds, as evidenced by brown discoloration of seedling and medium. Many inoculated seedlings showed severe leaf malformation which was also attributed to damage during inoculation.

Histochemical assays indicated GUS expression in some of the plants that survived (Figure 1). The rates of putative transformation, as seen by histochemical and fluorometric GUS expression, varied from 11% for QL27 inoculated with LBA4404 pBI121 to 3% for Tx430 inoculated with AGLO pKIWI105. This may be attributed to either host genotype differences, or differences in virulence between the two *Agrobacterium* strains. It is also important to point out that *Agrobacterium* strains with the binary plasmid have been shown to express GUS at low levels. This is not the case with the pKIWI105 construct which has never been observed to express GUS in the bacterium (R. Gardner, personal communication). Hence, the possibility cannot be discounted that some or all of the GUS expression observed in sorghum inoculated with LBA4404 pBI121 was of bacterial origin, and no gene transfer actually took place with this strain. Progeny tests were performed to confront this possibility, as it is unlikely that *Agrobacterium* cells would pass through the seed.

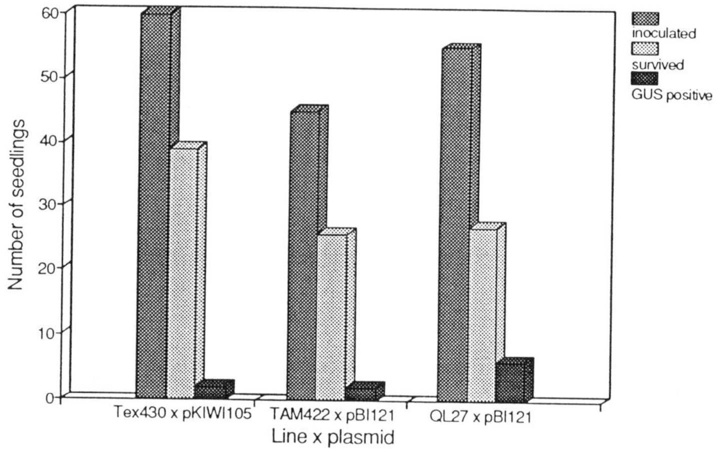

Figure 1. Seedling survival and histochemical GUS expression for 3 sorghum lines after inoculation with *Agrobacterium tumefaciens* with one of two binary vectors.

Five putative transformants and some of their progeny have been followed with respect to GUS expression. GUS expression was monitored throughout the growth of the plant, and only two, QL2 and Tm2, continued to express GUS in every leaf tested, including the flag leaf. We believe this to be a reflection of chimaeric transformation, which is not an unexpected result considering the differentiated, multicellular nature of the target tissue. After the booting stage most plants produced secondary tillers, and for plant Tm2 GUS expression was observed in all five tillers.

Seed set was generally poor. While this may be due in part to the *in vitro* germination and inoculation process, other possible contributing factors include low temperatures in the containment glasshouse during the winter when these plants were grown, and to some extent the effect of regular leaf harvest for histochemical GUS assays and DNA extraction. Histochemical GUS assays of the progenies indicated that stable incorporation of the T-DNA was achieved in some of the tillers (Table 1).

Southern blot hybridizations demonstrated incorporation of the GUS gene into high molecular weight DNA of transgenic tobacco and putative transgenic sorghum (data not presented). This is further evidence of incorporation of transgenes into the sorghum genome. There was no hybridization with negative control sorghum lanes. Due to problems with restriction enzyme purity the restriction digests of sorghum genomic DNA revealed a smear rather than any specific banding pattern. This was also evident in the tobacco positive transformed control.

Methylation of transgenes could explain the non-Mendelian segregation ratios for GUS expression. Chimaerism of primary transformants would also explain the ratios observed. Indeed, GUS expression patterns in vegetative parts of primary transformants was chimaeric in all but two cases. There was no GUS expression from the progeny of the main inflorescence, whereas there was in secondary tillers (Table 1).

Table 1. Segregation ratios for histochemical GUS expression in selfed progenies of putative primary transformant

Primary transformant	Tillers GUS+	Tiller	GUS+:GUS
QL1	2/8	1	0:28
QL2	4/11	2	5:11
QL3	1/9		
Tm1	3/10	1	0:30
		2	12:11
		3	16:9
		4	10:10
Tm2	5/5	2	11:14
		3	17:7
		4	15:5
		5	18:5

This work demonstrates that sorghum is within the host-range of *Agrobacterium* for *in vitro* transformation. The following evidence would appear to lend weight to our hypothesis:

1. Endogenous GUS expression has never been observed in untransformed control sorghum plants within 48 h of the assay
2. GUS expression was observed in plants inoculated with the vector pKIWI105 which has never been found to express GUS in the bacterium
3. GUS expression was predominantly chimaeric, which is to be expected with the needle inoculation of differentiated plant material
4. Progenies from GUS positive tillers were shown to segregate for GUS expression, although segregation ratios were not as expected.
5. DNA sequences homologous to a GUS probe was found from high molecular weight DNA in primary transformants and their progeny.

6. Mendelian segregation for GUS expression was observed in the progeny of one primary transformant.

The genetic transformation method reported here has a number of advantages over other alternatives. Most transformation protocols require at least some tissue culture step, usually with the need to differentiate cell cultures for plant regeneration. By utilising the natural germination process, this protocol avoids potential problems of genotype specificity and somaclonal variation. Regeneration from sorghum protoplasts is currently possible at very low frequency (Wei and Xu, 1990), and there are genotype-specificity problems associated with embryogenic cell cultures, hence it may be difficult to develop electroporation and microprojectile-based transformation protocols for sorghum. Our laboratory is currently developing a micro-projectile based approach for genetic transformation of embryogenic callus cultures.

Work is currently in progress to repeat these experiments on a range of plant materials and to attempt to improve the efficiency of the system. Whereas it may prove difficult to overcome the problems of high seedling mortality associated with the inoculation procedure, it is envisaged that efficiencies could be improved with the use of an effective selectable marker gene. Experiments indicate that phosphinothricin and hygromycin may be suitable selective agents (data not presented). Expression of both selectable marker and reporter genes will be tested under the control of cereal regulatory elements.

Acknowledgments

We thank the Australian International Development Assistance Bureau for providing a postgraduate scholarship for R. Chikwamba.

REFERENCES

Battraw, M. and Hall, T.C. (1991). Stable transformation of *Sorghum bicolor* protoplasts with chimeric neomycin phospotransferase II and β-glucuronidase genes. *Theoretical and Applied Genetics.* 82:161–168.

Bhaskaran, S. and Smith, R.H. (1988). Enhanced somatic embryogenesis in *Sorghum bicolor* from shoot tip culture. *In Vitro Cellular and Developmental Biology.* 24:65–70.

Bhaskaran, S. Smith, R.H. Paliwal, S. and Schertz, K.F. (1987). Somaclonal variation from *Sorghum bicolor* cell culture. *Plant Cell Tissue and Organ Culture.* 9:189–196.

Brettel, R.I.S. Wernicke, W. and Thomas, E. (1980). Embryogenesis from cultured immature inflorescences of *Sorghum bicolor. Protoplasma.* 104:141–148.

Chan, M.T., Chang, H.H., Ho, S.L., Tong, W.F. and Yu, S.M. (1993). *Agrobacterium*-mediated production of transgenic rice plants expressing a chimeric α-amylase promoter/β-glucuronidase gene. *Plant Molecular Biology.* 22:491–506.

deBlock, M. Botterman, J. Vandeweile, M. Dockx, J. Thoen, C. Gossele, V. Rao Movva, N. Thompson, C. Van Montagu, M. and Leemans, J. (1987). Engineering herbicide resistance in plants by expression of a detoxifying enzyme.*EMBO Journal.* 6:2513–2518.

Dellaporta, S.L., Wood, J. and Hicks, J.B. (1983). A plant DNA minipreparation: Version II. *Plant Molecular Biology Reporter.* 1:19–21.

Dunstan, D.I. Short, K.C. Dhaliwal, H. and Thomas, E. (1979). Further studies on plantlet production from cultured tissues of *Sorghum bicolor. Protoplasma.* 101:365–361.

Fromm, M.E. Morrish, F. Armstrong, C. Williams, R. Thomas, J. and Klein, T.M. (1990). Inheritance and expression in the progeny of transgenic maize plants. *Bio/Technology.* 8:833–839.

Feinberg, A.P. and Vogelstein, B. (1983). A technique for radiolabelling DNA restriction endonuclease fragments to high specific activity. *Analytical Biochemistry* 132:6–13.

Gould, J. Devey, M. Hasegawa, O. Ulian, E.C. Petersen, G. and Smith, R.H. (1991). Transformation of *Zea mays* using *Agrobacterium tumefaciens* and the shoot apex. *Plant Physiology.* 95:426–434.

Hagio, T. Blowers, A.D. and Earle, E.D. (1991). Stable transformation of sorghum cell cultures after bombardment with DNA-coated microprojectiles. *Plant Cell Reports.* 10:260–264.

Jefferson, R.A. (1987). Assaying chimeric genes in plants: The GUS gene fusion system. *Plant Molecular Biology Reporter.* 5:387–405.

Masteller, V.J. and Holden, D.J. (1970). The growth of organ formation from callus tissue of sorghum. *Plant Physiology.* 45:360–364.

Purseglove, J.W. (1972). *Tropical Crops: Monocotyledons*, Longman Group Ltd, Harlow, Essex, UK.

Raineri, D.M. Bottino, P. Gordon, M.P. and Nester, E.W. (1990). *Agrobacterium*-mediated transformation of rice (*Oryza sativa* L.) *Bio/Technology* 8:33–38.

Rhodes, C.A., Pierce, D.A., Mettler, I.J., Mascarenhas, D. and Detmer, J.J. (1988). Genetically transformed maize plants from protoplasts. *Science.* 240:204–207.

Sambrook, J., Fritsch, E.F. and Maniatis, T. (1989). *Molecular Cloning: A Laboratory Manual. 2nd Edition.* Cold Spring Harbor Laboratory Press. New York.

Shimamoto, K. Terada, R. Izawa, T. and Fujimoto, H. (1989). Fertile transgenic rice plants regenerated from transformed protoplasts. *Nature.* 338:274–276.

Somers, D.A., Rines, H.W., Gu, W., Kaeppler, H.F. and Bushnell, W.R. (1992). Fertile, transgenicoat plants. *Bio/Technology.* 10: 1589–1594.

Vasil, V. Castillo, A.M. Fromm, M.E. and Vasil, I.K. (1992). Herbicide resistant fertile transgenic wheat plants obtained by microprojectile bombardment of regenerable embryogenic callus. *Bio/Technology.* 10:667–674.

Wei, Z. and Xu, Z. (1990). Regeneration of fertile plants from embryogenic suspension culture protoplasts of *Sorghum vulgare. Plant Cell Reports.* 8:51–53.

Wernicke, W. and Brettel, R.I.S. (1980). Somatic embryogenesis from *Sorghum bicolor* leaves. *Nature.* 287:138–139.

DEVELOPMENT OF PROMOTER SYSTEMS FOR THE EXPRESSION OF FOREIGN GENES IN TRANSGENIC CEREALS

D. McElroy,[1*] W. Zhang,[2] D. Xu,[2] B. Witrzens,[1] F. Gubler,[3] J. Jacobsen,[1] R. Wu,[2] R. I. S. Brettell,[1] and E. S. Dennis[1]

[1]CSIRO Division of Plant Industry
GPO Box 1600
Canberra, ACT 2601, Australia

[2]Department of Biochemistry
Molecular and Cell Biology
Cornell University
Ithaca, NY 14853

[3]Co-operative Research Centre for Plant Science
GPO Box 475
Canberra, ACT 2601, Australia

SUMMARY

Recent advances in monocot transformation technology have resulted in the routine production of transgenic plants for an increasing number of cereal species. With a view towards the improvement of cereal quality by genetic engineering attention is beginning to focus on the identification and utilisation of promoters that will be used to control the expression of agronomically important traits in transformed cereals.

INTRODUCTION

There are a number of key elements that compose a successful plant transformation system. These include a means to deliver DNA to those cells that are competent for both transformation and regeneration, a method for plant regeneration and a selection protocol

* Corresponding author: present address: USDA/ARS Plant Gene Expression Center, 800 Buchanan St., Albany, CA 94710..

that acts to increase the frequency of transgenic plant recovery. Equally important components of any transformation strategy are the molecular elements that act to regulate the correct spatial and/or temporal expression pattern of the introduced genes, both in the primary transgenic plants and in the field grown crops derived from them.

It is over ten years since methods were first developed for the genetic engineering of higher plants using either *Agrobacterium*-mediated transformation of intact tissues or direct gene transfer to isolated protoplasts (Hernalsteens et al., 1980; Otten et al., 1981, Potrykus, 1990). For reasons of physiological incompatibility, cereals did not prove readily amenable to *Agrobacterium*-mediated gene delivery and the initial focus of transgenic cereal technology involved the development of methods to regenerate plants following direct gene transfer to isolated protoplasts (Cocking and Davey, 1987). Rice has led the way in cereal transformation, with both the first report of plant regeneration from protoplasts (Fujimura et al., 1985), and the first report of fertile transgenic plants regenerated from transformed protoplasts (Shimamoto et al., 1989). Although the regeneration of maize (Prioli et al., 1989; Shillito et al., 1989) and wheat (Vasil et al., 1990) plants from protoplasts has been described, the regeneration of fertile transgenic plants from protoplasts of these (and other) cereal species has proven to be technically demanding (Rhodes et al., 1988) and alternative technologies for use in cereal transformation have been developed.

The use of microparticle bombardment mediated transformation of embryogenic tissue culture material, with the subsequent regeneration and selection of transgenic plants, has overcome the problems associated with the production of transgenic plants from cereal protoplasts. Using this novel technology, transgenic plants have been obtained from microparticle bombarded suspension cultures of rice (Christou et al., 1991) and maize (Fromm et al., 1990; Gordon-Kamm et al., 1990) callus cultures of oats (Somers et al., 1992), sugarcane (Bower and Birch, 1992) and wheat (Vasil et al., 1992) and immature zygotic embryos of rice (Li et al., 1993), maize (Koziel et al., 1993) and wheat (Kartha et al., 1992).

Attention is now turning towards the identification and utilisation of promoter systems that can act to regulate the spatial and/or temporal expression of foreign genes in transgenic cereals. Although such studies are not as far advanced as those for transgenic dicots, a number of steps have been taken towards the development of such regulatory systems for use in cereal transformation. These steps include the development of novel reporter systems, the isolation and evaluation of appropriate promoter elements and the elucidation of mechanisms that act to enhance or interfere with foreign gene expression in transgenic cereals. A number of general principles have emerged from such studies. The application of these principles, in combination with the isolation, characterisation and regulated expression of agronomically important traits in transgenic cereals, will provide a basis for the future improvement of cereal quality by genetic engineering.

REPORTER GENES USED IN CEREAL TRANSFORMATION

Reporter genes are used in cereal transformation for the analysis of promoter activities, for monitoring selection efficiency (in both transformed tissue and transgenic plants) and for following the inheritance of foreign genes in subsequent plant generations. The utility of different reporter genes in cereal transformation is a function of the properties of the respective protein products they encode (Jefferson and Wilson, 1991). The required properties of a good reporter gene are summarised in Table 1. The reporter gene should show low background activity in transgenic cereals and should not have any detrimental effects on plant metabolism. The reporter gene product should have only moderate stability *in vivo* so as to detect down-regulation of gene expression as well as gene activation. The reporter gene should come with an assay system that is non-destructive, quantitative, sensitive, versatile, simple to carry out and inexpensive. As can be seen from Table 1, none of the reporter systems currently used in cereal transformation have all these desired properties and the relative limitations of each system should be borne in mind when utilising an individual reporter gene.

β-GLUCURONIDASE

The β-glucuronidase (*gus*) gene (Jefferson et al., 1987), encoded by the *uidA* locus of *E. coli*, is by far the most popular reporter gene used in cereal transformation. β-glucuronidase catalyses the hydrolysis and cleavage of a wide range of fluorometric and histochemical β-glucuronide substrates. The activity of the GUS enzyme can be easily and sensitively assayed in plants, the expression of *gus* gene fusions can be quantified by fluorometric assay, and histochemical analysis can be used to localise gene activity in transgenic tissues (Jefferson, 1987). However, *gus* gene expression assays are destructive and the GUS protein shows high *in vivo* stability, leading to problems when used to monitor gene inacti-

Table 1. Summary of Reporter Genes Used in Cereal Transformation

Properties	β-glucuronidase	Luciferase	Anthocyanin Regulators
Source	*E. coli* (Jefferson et al., 1987)	Firefly (de Wet et al., 1987)	Maize (Ludwig et al., 1990)
Background activity in plants	Low (some reports due to bacterial contaminants)	Low	Low–Moderate (depending upon species/ tissues)
Nature of Assay	Destructive	Destructive/ non-destructive	Non-destructive
Enzyme stability	High	Low	Low
Sensitivity of assay	High	Moderate	Low
Simplicity of assay	Good	Poor	Good
Quantitative nature of assay	Good	Good	Poor
Versatility of Assay (quantitative, histochemical, etc.)	Good	Poor	Poor
Adverse effects on transgenic plant metabolism	Low	Low	Some (at high levels)
Relative cost of assay systems	Moderate	High (requires expensive detection equipment)	Low
Reference to use in cereal transformation	Rice (Zhang and Wu, 1988)	Maize (Klein et al., 1989)	Maize (Goff et al., 1990; Ludwig et al., 1990)

vation. Furthermore, as will be described below, a dependence on the use of this bacterial reporter gene to monitor the efficiency of cereal transformation protocols can often be misleading.

LUCIFERASE

An alternative to the β-glucuronidase reporter gene system is the firefly (*Photinus pyralis*) luciferase gene (deWet et al., 1987). Luciferase catalyses the oxidation of D(–)-luciferin in the presence of ATP to generate oxyluciferin and yellow-green light. The activity of luciferase gene fusions can be assayed in transformed cereal tissue non-destructively (Klein et al., 1989; Goff et al., 1990). Unfortunately, penetration of the luciferin substrate can be limiting in whole plant material (R. Birch, personal communication) and the detection equipment presently needed to monitor luciferase gene expression is relatively expensive. These limitations have tended to preclude the widespread use of luciferase reporter genes in transgenic cereal plants. However, luciferase genes are widely used as an internal standard with *gus* fusions constructed to study gene expression in transgenic cereals (Luehrsen and Walbot, 1991).

ANTHOCYANIN BIOSYNTHETIC PATHWAY GENES

A reporter system that does not require the application of external substrates for its detection is based upon the *C1*, *B* and *R* genes which code for *trans*-acting factors that regulate the anthocyanin biosynthetic pathway in maize seeds. The introduction of these regulatory genes, under the control of constitutive promoters, into cereal cells by microparticle bombardment induces cell autonomous pigmentation in both seed and non-seed tissues (Ludwig et al., 1990; Goff et al. 1990). The activity of these reporter genes can be quantified in living tissue by counting the number of pigmented cells following bombardment. Although there is some evidence to suggest that overexpression of these *trans*-acting factors in transformed cereal tissues can be debilitating (R. Birch, personal communication) there are reports which indicate that these genes might be useful visible markers for selecting stably transformed cells lines that can eventually give rise to transgenic plants (Wang et al., 1990). An alternative means of utilising anthocyanin reporter genes in cereal transformation involves the complementation of mutations in genes of the anthocyanin biosynthetic pathway, for example the expression of the barley dihydroflavonol reductase gene in anthocyanin-free barley mutants (Bowen et al., 1992).

THE USE OF PROMOTERS TO CONTROL FOREIGN GENE EXPRESSION IN TRANSGENIC CEREALS

In the area of gene regulation in transgenic cereals, it is the characterisation of constitutive and non-constitutive 5′ promoter elements that has advanced the most. However, it should be borne in mind that there are other non-promoter elements which also contribute to the control of gene expression in transgenic plants. Such non-promoter components include (among others) 3′ located regulatory elements (Deanet al., 1989; Dietrich et al., 1992) as well as those which act to regulate transcript termination (Hernandez et al., 1989; Ingelbrecht et al., 1989; Mogen et al., 1990), transcript stability (Pierce et al., 1987; Newman et al.,

1993), post-transcriptional modification (Callis et al., 1987) and/or translation efficiency (Murray et al., 1990; McElroy et al., 1991; Wada et al., 1991).

CONSTITUTIVE PROMOTERS

As can be seen from Table 2, the use of constitutive promoters in transgenic cereals has generally been restricted to the expression of resistance genes in various cereal transformation systems. In transient assays of *gus* reporter gene fusion constructs it has been found that the constitutive promoters commonly used in cereal transformation show differences in their relative activity in monocot cells (McElroy et al., 1991). Therefore, it has been assumed that the relative level of resistance gene expression in transformed cereal cells can be controlled by changing the origin of its constitutive promoter.

The constitutive promoter of the cauliflower mosaic virus 35S RNA transcript (*35S*) has been used extensively in dicot transformation (Benfey and Chua, 1990). The *35S* promoter has been used to express selectable marker genes in transformed rice (Shimamoto et al., 1989; Christou et al., 1991; Cao et al., 1992; Meijer et al., 1991; Li et al., 1993) and maize (Rhodes et al., 1988; Gordon-Kamm et al., 1990). However, the *35S* promoter shows relatively low activity in transient assays of monocot cells transformed with *gus* fusion genes (McElroy et al., 1990, 1991; Last et al., 1991; Peterhans et al., 1991) By extrapolation it has been proposed, in our view erroneously, that improvements in selection efficiencies could be obtained by increasing the relative level of selectable marker gene expression in transformed cereal cells.

A number of strategies have been employed to increase gene expression in monocot cells. The incorporation of an intron into the foreign gene transcription unit has been found to elevate mRNA abundance and increase gene expression in transformed cereal cells (Callis et al., 1987; Vasil et al., 1989; Luersen and Walbot, 1991; McElroy et al., 1991). The maize alcohol dehydrogenase 1 gene (*Adh1*) first intron in a *35S* promoter/*Adh1* intron 1 combination has been used to control *bar* gene expression in transformed maize (Fromm et al., 1990), oat (Somers et al., 1992) and wheat (Vasil et al., 1992). An alternative strategy to enhance gene expression has been to modify a monocot promoter for high level constitutive activity in cereal cells. An example of this strategy is the modified maize *Adh1* sequence in the Emu promoter (Last et al., 1991) that has been used to control kanamycin resistance gene expression in transformed rice (Chamberlain et al., 1994) and sugarcane (Bower and Birch 1992). Finally, efforts have been made to isolate monocot promoters which naturally show high level constitutive activity in cereal cells. Examples of such sequences include the rice actin (*Act1*) and maize ubiquitin (*Ubi1*) promoters which have been used for *bar* gene expression in transgenic rice (Cao et al., 1992; Toki et al., 1992).

With the trend in cereal transformation systems moving towards microparticle bombardment of intact tissues with *bar* fusion genes (see below, Table 2) one might imagine that overexpression of such a herbicide detoxification gene would only lead to an increase in the frequency of cross-protection of untransformed cells by their transformed neighbours. Such cross-protection would result in an increase in the occurrence of non-transformed or chimaeric plants. This view is supported by the reported comparable efficiency of *35S-bar* and *Act1-bar* marker genes in the selection of transgenic plants from microparticle bombarded rice suspension culture cells (Cao et al., 1992). Therefore, the use of promoters with low to moderate levels of activity in monocot cells might provide an optimal strategy for the use of such detoxification genes in the selection of transgenic plants regenerated from intact cereal tissues. Alternatively, the development of selection systems based upon the over-expression

Table 2. Summary of Constitutive Promoters Used in Transgenic Cereals

Promotor	Use in Transgenic Cereals	Coding Region Expressed[a]	
35S Cauliflower Mosaic Virus 35S RNA transcript (Guilley et al., 1982)	Rice	*gus*	(Battraw and Hall, 1990; Teal and Shimamoto, 1990)
		nptII	(Toriyama et al., 1988; Zhang et al., 1988)
		hpt	(Shimamoto et al., 1989)
		bar	(Christou et al., 1991)
		als	(Li et al., 1992)
		dhfr	(Meijer et al., 1992)
	Maize	*nptII*	(Rhodes et al., 1988)
		bar	(Gordon-Kamm et al., 1990)
	Fescue	*hpt*	(Wang et al., 1992)
35S - Adh1 intron 1 Cauliflower Mosaic Virus 35S promoter and first intron of maize alcohol dehydrogenase 1 gene (Callis et al., 1987)	Maize	*bar*	(Fromm et al., 1990)
	Oats	*bar*	(Somers et al., 1992)
	Wheat	*bar*	(Vasil et al., 1992)
Emu Modified maize alcohol dehydrogenase 1 promoter and first intron (Last et al., 1991)	Rice	*nptII*	(Chamberlain et al., 1994)
	Sugarcane	*nptII*	(Bower and Birch, 1992)
Act1 - Act1 intron 1 Rice actin 1 gene (McElroy et al., 1990b)	Rice	*gus*	(Zhang et al., 1991)
		bar	(Cao et al., 1992)
Ubi1 - Ubi1 intron 1 Maize ubiquitin 1 gene (Christensen et al., 1992)	Rice	*bar*	(Toki et al., 1992)
	Wheat	*bar*	(Weeks et al., 1993)

[a] Abbreviations used: *gus*, β-glucuronidase gene; *nptII*, neomycin phosphotransferase II gene; *hpt*, hygromycin phosphotransferase gene; *bar*, phosphinothricin acetyl transferase gene; *als*, acetolactate synthase gene; *dhfr*, dihydrofolate reductase gene.

of novel herbicide tolerance genes in intact tissues might alleviate the problems of cross-protection that are associated with the activity of such detoxification genes. This is not to suggest that constitutive promoters and enhancement strategies designed to drive high level gene expression in monocots will not find a use in cereal transformation.

The uses of such promoters will include : the expression of suitable reporter genes to monitor the efficiency of transformation and selection protocols; the constitutive repression of endogenous (Shimada et al., 1993) and/or pathogenic gene expression through antisense technologies (Bejarano and Lichtenstein., 1992); the overproduction of biomolecules in transgenic plants and the overexpression and assaying of disease resistance genes in transgenic cereals prior to the subsequent employment of non-constitutive gene expression strategies (Lamb et al., 1992; Wilson, 1993).

TISSUE-SPECIFIC PROMOTERS

Advances in cereal transformation technology have tended to lag behind developments in dicot transformation systems. Furthermore, progress in dicot gene isolation and analysis has, in many cases, lead similar developments with homologous monocot genes. These trends have had two major effects on the study of foreign gene expression in transgenic cereals. Firstly, there has been an increasing availability of tissue-specific dicot promoters for the regulated expression of foreign genes in monocot plants, although it has (up until recently) been technically difficult to study the behaviour of such dicot promoters in transgenic cereals. Secondly, there has been a tendency to study monocot gene expression in either transgenic dicots or transient assays of transformed cereal cells such as protoplasts (Callis et al., 1987; McElroy et al., 1990) or microparticle bombarded tissues (Klein et al., 1989; Ludwig et al., 1990; Luehrsen and Walbot, 1991). However, there is evidence to suggest that promoter and other regulatory elements from monocot species are not always correctly controlled in either transgenic dicot plants or in monocot transient assay systems (Sakamoto et al., 1991; Zheng et al., 1993).

Recent progress in rice transformation has enabled researchers to begin to study the regulation of gene expression in transgenic cereal tissues using *gus* reporter gene fusions (Table 3). Whether the promoters used in such studies have originated from monocots (Zhang et al., 1988; Kyozuka et al., 1991, 1993; Tada et al., 1991; Terada et al., 1993; Zheng et al., 1993; F. Gubler, J. Jacobsen and B. Witrzens, unpublished result) dicots (Kyozuka et al., 1993; Xu et al., 1993) or plant pathogens (Matsuki et al. 1989, Bhattacharyya-Pakrasi et al. 1993), it has been found that inclusion of the promoter region alone is enough to give the expected pattern of reporter gene expression in transgenic rice. A comparison of the rice and tomato *rbcS* promoters (Kyozuka et al., 1993) has shown that monocot promoters can show higher activity than their dicot homologues in transgenic rice. Finally, the intron mediated enhancement of gene expression in monocot cells, while increasing the activity of the (dicot) *pinII* promoter in transgenic rice did not alter its pattern of regulation or expression (Xu et al., 1993). These findings should be borne in mind when planning the regulated expression of foreign genes in transgenic cereals

INACTIVATION OF FOREIGN GENE EXPRESSION IN TRANSGENIC CEREALS

There are a number of conditions that can act to interfere with gene expression in transgenic plants. These conditions include: poor recognition of promoter elements and differences in transcription signals between the source of the promoter and the transformed material (Peterhans et al., 1991); inefficient mRNA termination, polyadenylation, processing and/or stability (Callis et al., 1987; Pierce et al., 1987; Hernandez et al., 1989; Ingelbrecht et al., 1989; Mogen et al., 1990; McElroy et al., 1990); inefficient translation due to low rates of translation initiation and/or inappropriate codon usage (Murray et al., 1989; McElroy et al., 1991; Wada et al., 1991); poor design of expression vectors which might contain multiple repeated regions leading to potential gene inactivation by recombination, methylation and/or co-suppression, or multiple genes with opposing orientations leading to transcriptional interference and/or anti-sense mediated gene inactivation (Paszty and Lurquin, 1990; Ingelbrecht et al., 1991). All of these potential problems can be obviated by appropriate expression vector design. However, as transgenic cereals make their way from the laboratory to the field there are other endogenous mechanisms that can act to interfere with the expression of their foreign genes, leading to the non-Mendelian inheritance of introduced traits.

Table 3. Summary of Tissue-Specific Promoter Activities in Transgenic Rice using *gus* Reporter Gene Fusion Constructs

Promoter	Pattern of Promoter-*gus* Fusion Gene Expression in Transgenic Rice
Adh1 Maize alcohol dehydrogenase 1 gene (Dennis et al., 1984)	Constitutive in root caps, anthers, filaments, pollen, scutellum, endosperm and embryo shoot and root. Anaerobically induced in roots. (Zhang and Wu, 1988; Kyozuka et al., 1991)
rolC Open reading frame 12 (ORF12) of the *Agrobacterium rhizogenes* Ri plasmid TL-DNA region (Schmulling et al., 1988)	Vascular tissue and embryogenic tissue. (Matsuki et al., 1989)
LHCP Rice light harvesting chlorophyll a/b-binding gene of photosystem II (Matsuoka et al., 1990)	Light inducible in leaves, stems and floral organs. (Tada et al., 1991)
rbcS Rice and tomato ribulose-1,5-bisphosphate carboxylase/oxygenase small subunit gene (Sugita et al., 1987; Xie et al., 1987)	Light inducible in mesophyll cells. (Kyozuka et al., 1993)
His3 Wheat histone 3 gene (Tabata et al., 1984)	Root cells undergoing cell division (Terada et al., 1993)
PinII Potato wound-inducible II gene (Thornburg et al., 1987)	Systemic induction by wounding, methyl jasmonate and abscisic acid. (Xu et al., 1993)
RTBV Rice tungro bacilliform virus major transcript gene (Qu et al., 1991)	Leaf phloem tissue. (Bhattacharyya-Pakrasi et al. 1993)
Gt1 Rice glutelin 1 gene (Leisy et al., 1989)	Developing seed endosperm tissue. (Zheng et al., 1993)
Amy-pHV19 Barley high pI a-amylase gene 19 (Jacobsen and Close, 1991)	Scutellum and aleurone cells in germinating seed (F. Gubler, J. Jacobsen and B. Witrzens, unpublished result)

GENE INACTIVATION BY METHYLATION AND CO-SUPPRESSION

As monocot transformation technologies improve and an increasing number of transgenic cereals enter the field, there are a growing number of reports describing non-Mendelian inheritance of the introduced genes and the inactivation of foreign gene expression in the progenies of transformed cereal plants. These reports of gene inactivation have, to date, involved the loss of reporter gene expression under laboratory conditions and/or the loss of selectable marker gene expression under non-selective field environment. Such non-Mendelian inheritance and loss of gene expression would appear to be independent of both the cereal species transformed and the nature of the introduced genes (Meijer et al.,

1991; Davey et al., 1991; Walter et al., 1992; Schuh et al., 1993). While both copy number (Stockhaus et al., 1987) and the position of the integrated gene(s) (Jones et al., 1985) can influence gene expression in transgenic plants, other explanations have been invoked to account for the non-Mendelian inheritance of introduced genes and for the observation that some plants do not express all copies of their integrated foreign genes (Deroles and Gardner, 1988; Matzke et al., 1989; Scheid et al., 1991).

The most common DNA modification in plant cells is cytosine methylation at CG dinucleotides and CNG (where N can be any base) trinucleotides. Methylcytosine can interfere with protein-DNA interactions. Such methylation induced DNA modifications may be a normal part of the control of gene expression in plant cells. The regulation of seed storage and photosynthetic genes, the response of plant genes to environmental stimuli and the control of transposable elements in plants cells are all influenced by DNA methylation (Finnegan et al., 1993). Alterations in phenotypes are commonly observed in plants regenerated from cultured cells and tissues and there are an increasing number of reports in which both stable and reversible genetic alterations in both rice (Brown et al., 1990) and maize (Brettell and Dennis, 1991) have been linked to changes in patterns of DNA methylation.

It has been shown that methylation can influence *35S - gus* reporter gene expression intransgenic plants of tobacco (Weber and Graessmann, 1989) and rice (Meijer et al., 1991; R.I.S. Brettell, unpublished result) and that in both cases demethylation induced by 5-azacytidine is sufficient to reactivate *gus* gene expression (Weber et al., 1990; Meijer et al., 1991; R.I.S. Brettell, unpublished result). The recognition and inactivation of such introduced genes might be regarded as a defence mechanism to protect plants against the expression of potentially deleterious foreign DNAs (Doerfler, 1991). An immediate conclusion of these studies is that it might often be misleading to depend on the use of non-selectable reporter genes when following the efficiency of cereal transformation protocols.

In cereal transformation, whether using gene delivery by direct transfer to protoplasts or microparticle bombardment of intact plant cells, there is a tendency towards the integration of multiple copies of the introduced genes (Gordon-Kamm et al., 1990; Christou et al., 1991; Somers et al., 1992; Tada et al., 1991; Zhang et al., 1991; Cao et al., 1992; Wang et al., 1992). The term co-suppression has been used to describe a situation in plant transformation in which multiple copies of a gene are co-ordinately suppressed (van der Krol et al., 1990). It has been suggested that methylation of sequences required for active gene expression could account for this phenomenon (Finnegan et al., 1993). In dicot transformation such co-suppression usually involves the participation of an introduced gene and its endogenous homologue. However, gene inactivation has been reported following the introduction of different *35S* fusion constructs into transgenic dicots (Matzke et al., 1989; Tinland et al., 1991) and it would not be difficult to imagine similar effects occurring between multiple copies of foreign genes introduced into transformed cereal cells. Whatever the mechanism, the inactivation of introduced genes will have important implications for the maintenance of novel phenotypes in transgenic cereal plants.

FUTURE PROSPECTS

The immediate products of current efforts in monocot transformation will be the engineering of herbicide resistant cereals. A number of herbicides have been used as selective agents in cereal transformation (Wilmink and Dons, 1993) However, with the notable exception of sulfonylurea resistance in transgenic rice (Li et al., 1992), resistance to phosphinothricin (PPT) based herbicides using the *bar* gene from *Streptomyces hygroscopicus*

(Thompson et al., 1987) is fast becoming the method of choice for the selection of fertile transgenic cereals. PPT is a glutamate analogue which inhibits glutamine synthase leading to an accumulation of ammonia and eventual cell death (Tachibana et al., 1986a,b). The *bar* gene codes for phosphinothricin-N-acetyltransferase (PAT) which inactivates PPT by acetylation. In cereal transformation, PPT has an advantage over antibiotic agents in that it can be used to select resistant plants by addition to the tissue culture media or by spraying full-grown plants. PPT has proven to be effective in the microparticle bombardment mediated transformation of maize (Fromm et al., 1990; Gordon-Kamm et al., 1990), oats (Somers et al., 1992), rice (Christou et al., 1991; Cao et al., 1992; Toki et al., 1992) and wheat (Vasil et al., 1992, Weeks et al., 1993).

In the medium term, targets in cereal biotechnology will include (amongst many others): the engineering of disease resistance (Bejarano and Lichtenstein, 1992; Lamb et al., 1992; Oppenheim and Chet, 1992; Wilson, 1993); the modification of cereal grain quality (Visser and Jacobsen, 1993); and the development of gene tagging in transgenic cereals (Izawa et al., 1991; Murai et al., 1991; Shimamoto et al., 1993). Most of these applications will require a specific spatial and/or temporal control of gene expression. Results from promoter studies tend to suggest that (assuming no morphological or anatomical limitation) both monocot and dicot promoters maintain their correct pattern of activity in transgenic cereals. Furthermore, although monocot promoters might show higher levels of activity in cereal cells than their dicot homologues, the relatively low level of dicot promoter activity can be subjected to intron mediated enhancement without losing their pattern of activity.

For reasons which might be related to difference in the biochemistry, physiology and/or morphology between monocots and dicots, as well as to obviate potential problems with the use of non-cereal genetic elements in transgenic cereals, the isolation and utilisation of monocot promoters will continue to be a rewarding area of research activity. In this respect it is worth noting that the first report of a field trial of transgenic maize expressing an agronomically important trait involved the expression of insecticidal *Bacillus thuringiensis* resistance genes under the control of leaf and pollen specific promoters from maize to target gene expression to those tissues that are most susceptible to attack by the European corn borer (Koziel et al., 1993). Whatever the future may hold, an understanding of gene expression in transgenic cereals will continue to pay high dividends in this ever expanding area of biotechnology.

Acknowledgments

The authors wish to thank Dr. Mick Graham for critical reading of the manuscript and for his many helpful suggestions. DM was supported by a postdoctoral research fellowship from The Australian Grains Research and Development Corporation, Grant CSP15.

REFERENCES

Battraw, M.J. and Hall, T.C. (1990) Histochemical analysis of CaMV 35S promoter—β-glucuronidase gene expression in transgenic rice plants. *Plant Mol. Biol.* 15, 527–538.

Battraw, M. and Hall, T.C. (1992) Expression of a chimeric neomycin phosphotransferase II gene in first and second generation transgenic rice plants. *Plant Science* 86, 191–202.

Bejarano, E.R. and Lichtenstein, C.P. (1992) Prospects for engineering virus resistance in plants with antisense RNA. *Tibtech* 10, 383–388.

Benfey, P.N. and Chua, N.-H. (1990) The cauliflower mosaic virus 35S promoter: combinatorial regulation of transcription in plants. *Science* 250, 959–966.

Bhattacharyya-Pakrasi, M., Peng, J., Elmer, J.S., Laco, G., Shen, P., Kaniewska, M.B., Kononowicz, H., Wen, F., Hodges, T.K. and Beachy, R.N. (1993) Specificity of a promoter from the rice tungro bacilliform virus for expression in phloem tissue. *Plant J.* 4, 71–79.

Bowen, B. (1992) Anthocyanin genes as visual markers in transformed maize tissue. In: GUS protocols: using the GUS gene as a reporter of gene expression. Ed. S. Gallagher. Academic Press, NY, USA. pp163–179.

Bower R. and Birch R.G. (1992) Transgenic sugarcane plants via microparticle bombardment. *Plant J.* 2, 409–416.

Brettell , R.I.S. and Dennis, E.S. (1991) Reactivation of a silent *Ac* following tissue culture is associated with heritable alterations in its methylation pattern. *Mol. Gen. Genet.* 229, 365–372.

Brown, P.T.H., Kyozuka, J., Sukekiko, Y., Kimura, Y., Shimamoto, K. and Lorz, H. (1990) Molecular changes in protoplast-derived rice plants. *Mol. Gen. Genet.* 223, 324–328.

Callis, J., Fromm, M. and Walbot, V. (1987) Introns increase gene expression in cultured maize cells. Genes and Devel. 1, 1183–1200.

Cao, J., Duan, X., McElroy, D. and Wu, R. (1992) Regeneration of herbicide resistant transgenic rice plants following microparticle-mediated transformation of suspension culture cells. *Plant Cell Rep.* 11, 586–591.

Chamberlain, D.A., Brettell, R.I.S., Last, D.I., Witrzens, B., McElroy, D. and Dennis, E.S. (1994) The use of the Emu promoter with antibiotic and herbicide resistance genes for the selection of transgenic wheat callus and rice plants. Aust. J. Plant. Physiol. 21, 95–112.

Christensen, A.H., Sharrock, R.A. and Quail, P.H. (1992) Maize polyubiquitin genes: structure, thermal perturbation of expression and transcript splicing, and promoter activity following transfer to protoplasts by electroporation. *Plant Mol. Biol.* 18, 675–689.

Christou, P., Ford, T. and Kofron, M. (1991) Production of transgenic rice (*Oryza sativa* L.) plants from agronomically important indica and japonica varieties via electric discharge particle acceleration of exogenous DNA into immature zygotic embryos. *Bio/Technology* 9, 957–962.

Cocking ,E.C and Davey, M.R. (1987) Gene transfer in cereals. *Science* 236, 1259–1262.

Davey, M.R., Kothari, S.L., Zhang, H., Rech, E.L., Cocking, E.C. and Lynch, P.T. (1991) Transgenic rice: characterization of protoplast-derived plants and their seed progeny. *J. Exp. Bot.* 42, 1159–1169.

Dean, C., Favreau, M., Bond-Nutter, D., Bedbrook, J., and Dunsmuir, P. ((1989) Sequences downstream of translation start regulate quantitative expression in two petunia genes. *Plant Cell* 1, 201–208.

de Wet, J.R., Wood, K.V., DeLuca, M., Helinski, D.R. and Subramani, S. (1987) Firefly luciferase gene: structure and expression in mammalian cells. *Mol. Cell. Biol.* 7, 725–737.

Dennis, E.S., Gerlach, W.L., Pryor, A.J., Bennetzen, J.L., Inglis, A., Llewellyn, C., Sachs, M.M., Ferl, R.J. and Peacock, W.J. (1984) Molecular analysis of the alcohol dehydrogenase (*Adh1*) gene of maize. *Nucl. Acids Res.* 12, 3983–4000.

Deroles, S.C. and Gardner, R.C. (1988) Expression and inheritance of kanamycin resistance in large numbers of transgenic petunias generated by *Agrobacterium*-mediated transformation. *Plant Mol. Biol.* 11, 355–364

Dietrich, R.A., Radke S.E. and Harada J.J. (1992) Downstream DNA sequences are required to activate a gene expressed in the root cortex of embryos and seedlings. *Plant Cell* 4, 1371–1382.

Doerfler, W. (1991) Patterns of DNA methylation—evolutionary vestiges of foreign DNA inactivation as a host defence mechanism. *Biol. Chem.* 372, 557–564.

Finnegan, E.J., Brettell, R.I.S. and Dennis, E.S. (1993) The role of DNA methylation in the regulation of plant gene expression. In: *DNA Methylation: Molecular Biology and Biological Significance.* Ed. J.P. Jost and H.P. Saluz. Birkhäuser Verlag, Basel, Switzerland. pp 218–261.

Fromm, M.E., Morrish, F., Armstrong, C., Williams, R., Thomas, J. and Klein, T.M. (1990) Inheritance and expression of chimaeric genes in the progeny of transgenic maize plants. *Bio/Technology* 8, 833–844.

Fujimura, T., Sakurai, M., Akagi, H., Negishi, T. and Hirose, A. (1985) Regeneration of rice plants from protoplasts. *Plant Tissue Culture Lett.* 2, 74–75.

Goff, S.A., Klein, T.M., Roth, B.A., Fromm, M.E., Cone, K.C., Radicella, J.P. and Chandler, V.L. (1990) Transactivation of anthocyanin biosynthetic genes following transfer of *B* regulatory genes into maize tissues. *EMBO J.* 9, 2517–2522.

Gordon-Kamm, W.J., Spencer, T.M., Mangano, M.L., Adams, T.R., Daines, R.J., Start, W.G., O'Brien, J.V., Chambers, S.A., Adams, W.R., Willetts, N.G., Rice, T.B., Mackey, C.J., Krueger, R.W., Kausch, A.P. and Lemaux, P.G. (1990) Transformation of maize cells and regeneration of fertile transgenic plants. *Plant Cell* 2, 603–618.

Guilley, H., Dudley, R.K., Jonard, G., Balazs, E. and Richards, K.E. (1982) Transcription of cauliflower mosaic virus DNA: Detection of promoter sequences, and characterization of transcripts. *Cell* 30, 763–773.

Hernalsteens, J.P., Van Vliet, F., De Beuckeleer, M., Depicker, A., Engler, G., Lemmers, M., Holsters, M., Van Montagu, M. and Schell, J. (1980) The *Agrobacterium tumefaciens* Ti plasmid as a host vector system for introducing foreign DNA in plant cells. *Nature* 287, 654–656.

Hernandez, G., Cannon, F. and Cannon, M. (1989) The effect of presumptive polyadenylation signals on the expression of the CAT gene in transgenic tobacco. *Plant Cell Rep.* 8, 195–198.

Ingelbrecht, I.L.W., Herman, L.M.F., Dekeyser, R.A., Van Montague, M.C. and Depicker, A.G. (1989) Different 3′ end regions strongly influence the level of gene expression in plant cells. *Plant Cell* 1, 671–680.

Ingelbrecht, I., Breyne, P., Vancompernolle, K., Jacobs, A., Van Montague, M. and Depicker, A. (1991) Transcriptional interference in transgenic plants. *Gene* 109, 239–242.

Izawa, T., Miyazaki, C., Yamamoto, M., Tereda, R., Iida, S. and Shimamoto, K. (1991) Introduction and transposition of the maize transposable element *Ac* in rice (*Oryza sativa* L.). *Mol. Gen. Genet.* 227, 391–396.

Jacobsen, J.V. and Close T.J. (1991) Control of transient expression of chimaeric genes by gibberellic acid and abscisic acid in protoplasts prepared from mature barley aleurone layers. *Plant Mol. Biol.* 16, 713–724.

Jefferson, R.A. (1987) Assaying chimeric genes in plants: The GUS gene fusion system. *Plant Mol. Biol Rep.* 5, 387–405.

Jefferson, R.A., Kavanagh, T.A. and Bevan, M.W. (1987) GUS Fusions: β-glucuronidase as a sensitive and versatile gene fusion marker in higher plants. *EMBO J.* 6, 3901–3907.

Jefferson, R.A. and Wilson, K.J. (1991) The GUS gene fusion system. In: *Plant Molecular Biology Manual.* Kluwer Academic Publishers, pp1–33.

Jones, J.D., Dunsmuir, P. and Bedbrook, J. (1985) High level expression of introduced chimaeric genes in regenerated transformed plants. *EMBO J.* 4, 2411–2418.

Kartha, K.K., Chibbar, R.N., Nehra, N.S., Leung, N., Caswell, K., Baga, M., Mallard, C.S. and Steinhauer, L. (1992) Genetic engineering of wheat through microprojectile bombardment using immature zygotic embryos. *J. Cellular Biochem. Supp.* 16F, 198.

Klein, T.M., Roth, B.A. and Fromm, M.E. (1989) Regulation of anthocyanin biosynthetic genes introduced into intact maize tissues by microprojectiles. *Proc Natl. Acad. Sci.* USA 86, 6681–6685.

Koziel, M.G., Beland, G.L., Bowman, C., Carozzi, N.B., Crenshaw, R., Crossland, L., Dawson, J., Desai, N., Hill, M., Kadwell, S., Launis, K., Lewis, K., Maddox, D., McPherson, K., Meghji, M.R., Merlin, E., Rhodes, R., Warren, G.W., Wright, M. and Evola, S.V. (1993) Field performance of elite transgenic maize plants expressing an insecticidal protein derived from *Bacillus thuringiensis*. *Bio/Technology* 11, 194–200.

Kyozuka, J., Fujimoto, H., Izawa, T. and Shimamoto, K. (1991) Anaerobic induction and tissue-specific expression of maize *Adh1* promoter in transgenic rice plants and their progeny. *Mol. Gen. Genet.* 228, 40–48.

Kyozuka, J., McElroy, D., Hayakawa, T., Xie, Y., Wu, R. and Shimamoto, K. (1993) Light-regulated and cell specific expression of tomato *rbcS-gusA* and rice *rbcS-gusA* fusion genes in transgenic rice. *Plant Physiol.* in press.

Lamb, C.J., Ryals, J.A., Ward, E.R. and Dixon, R.A. (1992) Emerging strategies for enhancing crop resistance to microbial pathogens. *Bio/Technology* 10, 1436–1445.

Last, D.I., Brettell, R.I.S., Chamberlain, D.A., Chaudhury, A.M., Larkin, P.J., Marsh, E.L. Peacock, W.J. and Dennis, E.S. (1991) pEmu: An improved promoter for gene expression in cereal cells. *Theor. Appl. Genet.* 81, 581–588.

Leisy, D.J., Hnilo, J., Zhao, Y. and Okita, T.W. (1989) Expression of a rice glutelin promoter in transgenic tobacco. *Plant Mol. Biol.* 14,41–50.

Li, L., Qu, R., de Kochko, A., Fauquet, C. and Beachy, R.N. (1993) An improved rice transformation system using the biolistic method. *Plant Cell Rep.* 12, 250–255.

Li, Z., Hayashimoto, A. and Murai N. (1992) A sulfonylurea herbicide gene from *Arabidopsis thaliana* as a new selectable marker for production of fertile transgenic rice plants. *Plant Physiol.* 100, 662–668.

Ludwig, S.R., Bowen, B., Beach, L. and Wessler, S.R. (1990) A regulatory gene as a novel visible marker for maize transformation. *Science* 247, 449–450.

Luehrsen, K.R. and Walbot, V. (1991) Intron enhancement of gene expression and the splicing efficiency of introns in maize cells. *Mol. Gen. Genet.* 225, 81–93.

Matsuki, R., Onodera, H., Yamauchi, T. and Uchimiya, H. (1989) Tissue-specific expression of the *rolC* promoter of the Ri plasmid in transgenic rice plants. *Mol. Gen. Genet.* 220, 12–16.

Matsuoka, M. (1990) Classification and characterization of cDNA that encodes the light-harvesting chlorophyll a/b binding protein of photosystem II from rice. *Plant Cell Physiol.* 31, 519–526.

Matzke, M.A., Primig, M., Trnovsky, J. and Matzke, A.J.M. (1989) Reversible methylation and inactivation of marker genes in sequentially transformed tobacco plants. *EMBO J.* 8, 643–649.

McElroy, D., Zhang, W., Cao, J. and Wu, R. (1990a) Isolation of an efficient actin promoter for use in rice transformation. *Plant Cell* 2, 163–171.

McElroy, D., Rothenberg, M., Reece, K.S. and Wu, R. (1990b) Characterization of the rice (*Oryza sativa*) actin gene family. *Plant Mol. Biol.* 15, 257–268.

McElroy, D., Blowers, A.D., Jenes, B. and Wu, R. (1991) Construction of expression vectors based on the rice actin 1 (*Act1*) 5′ region for use in monocot transformation. *Mol. Gen. Genet.* 231, 150–160.

Meijer, E.G.M., Schilperoort, R.A., van Os-Ruygrot, P.E. and Hensgens, L.A.M. (1991) Transgenic rice cell lines and plants: expression of transferred chimeric genes. *Plant Mol. Biol.* 16, 807–820.

Mogen, B.D., MacDonald, M.H., Graybosch, R. and Hunt, A.G. (1990) Upstream sequences other than AAUAAA are required for efficient messenger RNA 3′-end formation in plants. *Plant Cell* 2, 1261–1272.

Murai, N., Li, Z., Kawagoe, Y. and Hayashimoto, A. (1991) Transposition of the maize *activator* element in transgenic rice plants. *Nucl. Acids Res.* 19, 617–622.

Murray, E.E., Lotzer, J. and Eberle, M. (1989) Codon usage in plant genes. *Nucl. Acids Res.* 17, 477–499.

Newman, T.C., Ohme-Takagi, M., Taylor, C.B. and Green, P.J. (1993) DST sequences, highly conserved among plant *SAUR* genes, target reporter transcripts for rapid decay in tobacco. *Plant Cell* 5, 701–714.

Oppenheim, A.B. and Chet, I. (1992) Cloned chitinase in fungal plant-pathogen control strategies. *Tibtech* 10, 392–394.

Otten, L., De Greve, H., Hernalsteens, J.P., Van Montagu, M., Schieder, O., Straub, J. and Schell, J. (1981) Mendelian transmission of genes introduced into plants by the Ti plasmid of *Agrobacterium tumefaciens*. *Mol. Gen. Genet.* 183, 209–213.

Paszty, C.J.R. and Lurquin, P.F. (1990.) Inhibition of transgene expression in plant protoplasts by the presence in cis of an opposing 3′-promoter. *Plant Science* 772, 69–79.

Peterhans, A., Datta, S.K., Datta, K., Goodall, G.J., Potrykus, I. and Paszkowski, J. (1991) Recognition efficiency of *Dicotyledoneae*-specific promoter and RNA processing signals in rice. *Mol. Gen. Genet.* 222, 361–368.

Pierce, D.A., Mettler, I.J., Lachmansingh, A.R., Pomeroy, L.M., Weck, E.A. and Mascarenhas, D. (1987) Effects of 35S leader modification on promoter activity. In: *Plant Gene Systems and Their Biology.* Eds. L. McIntosh and J. Key, Alan R. Liss, Inc, NY, USA. p301–310.

Potrykus, I. (1990) Gene transfer to plants: assessment and perspectives. *Physiol. Plantarum* 79, 125–134.

Prioli, L.M. and Sondahl, M.R. (1989) Plant regeneration and recovery of fertile plants from protoplasts of maize (*Zea mays* L.). *Bio/Technology* 7, 589–594.

Qu, R., Bhattacharyya, M., Laco, G.S., De Kochko, A., Subba Rao, B.L., Kaniewska, M.B., Elmer, J.S., Rochester, D.E., Smith, C.E. and Beachy, R.N. (1991) Characterization of the genome of rice tungro bacilliform virus: comparison with commelina yellow mottle virus and caulimovirises. *Virology* 185, 354–364.

Rhodes, C.A., Pierce, D.A., Mettler, I.J., Mascarenhas, D. and Detmer, J.J. (1988) Genetically transformed maize plants from protoplasts. *Science* 240, 204–207.

Sakamoto, M., Sanada, Y., Tagiri, A., Murakami, T., Ohashi, Y. and Matsuoka, M. (1991) Structure and characterization of a gene for light-harvesting Chl a/b binding protein from rice. *Plant and Cell Physiol.* 32, 385–394.

Scheid, O.M., Paszkowski, J. and Potrykus, I. (1991) Reversible inactivation of a transgene in *Arabidopsis thaliana*. *Mol. Gen. Genet.* 228, 104–112.

Schmulling, T., Schell, J. and Spena, A. (1988) Promoters of the *rolA*, *B*, and *C* genes of *Agrobacterium rhizogenes* are differentially regulated in transgenic plants. *Plant Cell* 1, 665–670.

Schuh, W., Nelson, M.R., Bigelow, D.M., Orum, T.V., Orth, C.E., Lynch, P.T., Eyles, P.S., Blackhall, N.W., Jones, J., Cocking, E.C. and Davey, M.R. (1993) The phenotypic characterization of R2 generation transgenic rice plants under field conditions. *Plant Sciences* 89, 69–79.

Shillito, R.D., Carswell, G.K., Johnson, C.M., DiMaio, J.J. and Harms, C.T. (1989) Regeneration of fertile plants from protoplasts of elite inbred maize. *Bio/Technology* 7, 581–587.

Shimada, H., Tada, Y., Kawasaki, T. and Fujimura, T. (1993) Antisense regulation of the rice *waxy* gene expression using a PCR-amplified fragment of the rice genome reduces the amylose content in grain starch. *Theor Appl Genet.* 86,665–672.

Shimamoto, K., Terada, R., Izawa, T. and Fujimoto, H. (1989) Fertile transgenic rice plants regenerated from transformed protoplasts. *Nature* 338, 274–276.

Shimamoto, K., Miyazaki, C., Hashimoto, H., Izawa, T., Itoh, K., Terada, R., Inagaki, Y. and Iida, S. (1993) *Trans*-activation and stable integration of the maize transposable element *Ds* cotransfected with the *Ac* transposase gene in transgenic rice plants. *Mol. Gen. Genet.* 239, 354–360.

Somers, D.A., Rines, H.W., Gu, W., Kaeppler, H.F. and Bushnell, W.R. (1992) Fertile, transgenic oat plants. *Bio/Technology* 10, 1589–1594.

Stockhaus, J., Eckes, P., Blau, A., Schell, J. and Willmitzer, L. (1987) Organ-specific and dosage-dependent expression of a leaf/stem specific gene from potato after tagging and transfer into potato and tobacco plants. *Nucl. Acids Res.* 15, 3479–3491.

Sugita, M., Manzara, T., Pichersky, E., Cashmore, A. and Gruissem, W. (1987) Genome organization, sequence analysis and expression of all five genes encoding the small subunit of ribulose-1,5-bisphosphate carboxylase/oxygenase from tomato. *Mol. Gen. Genet.* 209, 247–256.

Tabata, T., Fukasawa, M. and Iwabuchi, M. (1984) Nucleotide sequence and genomic organization of a wheat histone H3 gene. *Mol. Gen. Genet.* 196, 397–400.

Tachibana, K., Watanabe, T., Sekizawa, Y. and Takematsu, T. (1986a) Inhibition of glutamine synthase and quantitative changes of free amino acids in shoots of bialaphos-treated Japanese barnyard millet. *J. Pesticide Sci.* 11, 27–31.

Tachibana K., Watanabe T., Sekizawa Y. and Takematsu, T. (1986b) Accumulation of ammonia in plants treated with bialaphos. *J. Pesticide Sci.* 11, 33–37.

Tada, Y., Sakamoto, M., Matsuoka, M. and Fujimura, T. (1991) Expression of a monocot LHCP promoter in transgenic rice. *EMBO J.* 10, 1803–1808.

Tereda, R. and Shimamoto, K. (1990) Expression of CaMV35S-GUS gene in transgenic rice plants. *Mol. Gen. Genet.* 220, 389–392.

Terada, R., Nakayama, T., Iwabuchi, M. and Shimamoto, K. (1993) A wheat histone H3 promoter confers cell division-dependent and -independent expression of the *gusA* gene in transgenic rice plants. *Plant J.* 3, 241–252.

Thompson, C.J., Movva, N.R., Tizard, R., Crameri, R., Davies, J.E., Lauwereys, M and Botterman, J. (1987) Characterization of the herbicide-resistance gene bar from *Streptomyces hygroscopicus*. *EMBO J.* 9, 2519–2523.

Thornburg ,R.W., An G., Cleveland, T., Johnson, R. and Ryan, C.A. (1987) Wound-inducible expression of a potato inhibitor II-chloramphenicol acetyl transferase gene fusion in transgenic tobacco plants. *Proc. Natl. Acad. Sci. USA* 84, 744–748.

Tinland, B., Kares, C., Herrmann, A. and Otten, L. (1991) 35S-b-glucuronidase gene blocks biological effects of cotransformed *iaa* genes. *Plant Mol. Biol.* 16, 853–864.

Toki, S., Takamatsu, S., Nojiri, C., Ooba, S., Anzai, H., Iwata, M., Christensen, A.H., Quail, P.H. and Uchimiya, H. (1992) Expression of a maize ubiquitin gene promoter-bar chimeric gene in transgenic rice plants. *Plant Physiol.* 100, 1503–1507.

Toriyama, K., Arimoto, Y., Uchimiya, H. and Hinata, K. (1988) Transgenic rice plants after direct gene transfer into protoplasts. *Bio/Technology* 6, 1072–1074.

van den Krol, A.R., Mur, L.A., Beld, M., Mol, J.N.M. and Stuitje, A.R. (1990) Flavonoid genes in petunia: addition of a limited number of gene copies may lead to a suppression of gene expression. *Plant Cell* 2, 291–299.

Vasil, V., Clancy, M., Ferl, R., Vasil, I.K. and Hannah, L.C. (1989) Increased gene expression by the first intron of maize *shrunken-1* locus in grass species. *Plant Physiol* 91, 1575–1579.

Vasil, V., Redway, F.A. and Vasil, I.K. (1990) Regeneration of plants from embryogenic suspension culture protoplasts of wheat. *Bio/Technology* 8, 429–433.

Vasil, V., Castillo, A.M., Fromm, M.E. and Vasil, I.K. (1992) Herbicide resistant fertile transgenic wheat plants obtained by microparticle bombardment of regenerable embryogenic callus. *Bio/Technology* 10, 667–674.

Visser, R.G.F. and Jacobsen, E. (1993) Towards modifying plants for altered starch content and composition. *Tibtech* 11, 63–68.

Wada, K.-N., Wada, Y., Doi, H., Ishibashi, F., Gojobori, T. and Ikemura, T. (1991) Codon usage tabulated from the GenBank genetic sequence data. *Nucl. Acids Res.* 19, 1981–1986.

Waldron, C., Murphy, E.B., Roberts, J.L., Gustafson, G.D., Armour, S.L. and Malcolm, S.K. (1985) Resistance to hygromycin B, a new marker for plant transformation studies. *Plant Mol. Biol.* 5, 103–108.

Walters, D.A., Vetsch, C.S., Potts, D.E. and Lundquist, R.C. (1992) Transformation and inheritance of a hygromycin phosphotransferase gene in maize plants. *Plant Mol. Biol.* 18, 189–200.

Wang, X., Olsen, O. and Knudsen, S. (1993) Expression of the dihydroflavonol reductase gene in an anthocyanin-free barley mutant. Hereditas 119, 67–75.

Wang, Z-Y., Takamizo, T., Iglesias, V.A., Osusky, M., Nagel, J., Potrykus, I. and Spangenberg, G. (1992) Transgenic plants of tall fescue (*Festuca arundinacea* Schreb.) obtained by direct gene transfer to protoplasts. *Bio/Technology* 10, 691–696.

Weber, H. and Graessmann, A. (1989) Biological activity of hemimethylated and single stranded DNA after gene transfer into protoplasts. *FEBS Lett.* 253, 163–166.

Weber, H., Zeichmann, C. and Graessmann, A. (1990) In vitro DNA methylation inhibits gene expression in transgenic tobacco. *EMBO J.* 9, 4409–4415.

Weeks, J.T., Anderson, O.D. and Blechl, A.E. (1993) Rapid production of multiple independent lines of fertile transgenic wheat (*Triticum aestivum*) *Plant Physiol.* 102,1077–1084.

Wilmink, A. and Dons, J.J.M. (1993) Selective agents and marker genes for use in transformation of monocotyledenous plants. *Plant Mol. Biol. Rep.* 11, 165–185.

Wilson, T.M.A. (1993) Strategies to protect crop plants against viruses: Pathogen-derived resistance blossoms. *Proc. Natl. Acad. Sci. USA* 90, 3134–3141.

Xie, Y., Peng, Z., Cai, Y. and Wu, R. (1987) Cloning and analysis of histone 3 gene and rubisco small subunit gene of rice. *Sci. Sinica* (ser. B) 30, 706–718.

Xu, D., McElroy, D., Thornburg, R. and Wu, R. (1993) Systemic induction of a potato *pin2* promoter by wounding, methyl jasmonate, and abscisic acid in transgenic rice plants. *Plant Mol. Biol.* 22, 573–588.

Zhang, H.M., Yang, H., Rech, E.L., Golds, T.J., Davis, A.S., Mulligan, B.J., Cocking, E.C. and Davey, M.R. (1988) Transgenic rice plants produced by electroporation-mediated plasmid uptake into protoplasts. *Plant Cell Rep.* 7, 379–384.

Zhang, W. and Wu, R. (1988) Efficient regeneration of transgenic plants from rice protoplasts and correctly regulated expression of the foreign gene in the plants. *Theor. Appl. Genet.* 76, 835–840.

Zhang, W., McElroy, D. and Wu, R. (1991) Analysis of Rice *Act1* 5' region activity in transgenic rice plants. *Plant Cell* 3, 1155–1165.

Zheng, Z., Kawagoe, Y., Xiao, S., Okita, T., Hau, T.L., Lin, A. and Murai, N. (1993) 5′ distal and proximal cis-acting regulatory elements are required for developmental control of a rice seed storage protein *glutelin* gene. *Plant J.* 4, 357–366.

ANTHOCYANIN GENES AS VISUAL MARKERS FOR WHEAT TRANSFORMATION

S.K. Dhir, M.E. Pajeau, M.E. Frommn and J.E. Fry

Agriculture Group of Monsanto
700 Chesterfield Parkway North
St. Louis, Missouri 63198

INTRODUCTION

Regulatory genes of the maize anthocyanin biosynthetic pathway have proven useful as scorable markers for transformation because of their high sensitivity, ease of visualization, cell-autonomous expression and lack of requirement for exogenous substrates. This is particularly advantageous when using particle bombardment as a DNA delivery system, since cells expressing genes can be counted easily and unambiguously (Bowen 1992). Furthermore, the cells that have been damaged by microprojectile bombardment can still express GUS (giving false positive), but they will not accumulate anthocyanin since the latter requires an intact vacuole and coordinate expression of many genes. Expression of anthocyanin genes in plant tissues will be useful where false positive results have been reported due to intrinsic GUS-like activity in several plant parts (Hu et al. 1990) and GUS production by endophytic micro-organisms (Tor et al. 1992). The anthocyanin markers permit visualization of transgenic tissue from the beginning and throughout development without sacrificing the tissues. In addition, expression of anthocyanin genes in transformed tissues can provide a system for investigating the regulation of gene expression in plants, which is difficult with GUS (Benfey et al. 1990). Transient and stable expression of anthocyanin genes in various corn intact tissues (Klein et al. 1988; Ludwig et al. 1990; Wong et al. 1991; Bowen 1992; Dunder et al. 1993) and transient expression in wheat seedling tissues (Wong et al. 1991) has been reported previously. We have used anthocyanin as a visual marker for optimization of bombardment conditions using different target tissues (transiently) and evaluation of different selectable markers (stably) for wheat transformation by selecting the transformants visually in the early stage of development.

Improvement of Cereal Quality by Genetic Engineering, Edited by
Robert J. Henry and John A. Ronalds, Plenum Press, New York, 1994

MATERIAL AND METHODS

Plant Material

Immature embryos, approximately 0.8–1.5 mm in size (12–13 d after pollination) were aseptically excised from the kernels of different genotypes. Embryos were cultured on solidified MS (Murashige & Skoog's, 1962) medium supplemented with 2% sucrose, and 2 mg/l 2,4-dichlorophenoxy acetic acid (2,4-D) according to Redway et al. (1990). The resulting calli, which were embryogenic and nodular in morphology were selected and 2–3 mm diameter calli were arranged in petri dishes for bombardment.

Mustang suspension cells (obtained from W. C. Wang, Texas Tech University) were subcultured 3 days prior to bombardment. Suspension cells (approximately 0.5 ml packed cell volume, PCV) were plated onto 5.5 cm filters (Baxter, F2217-55). These filters were transferred to petri dishes containing 3 layers of filter paper.

Microprojectile Bombardment

Prior to bombardment plasmid DNA BC17 (anthocyanin genes under 35S promoter) and pMON19574 (E35S/CP4/NOS E35S/GOX/ NOS), or pMON19335 (E35S/DHFR/OS), or pMON19477 (E35S/BAR/NOS) with a monocot intron to enhance gene expression (Callis et al. 1987) were mixed and precipitated onto tungsten particles. Cells were bombarded using a Biolistic bombardment device as described by Klein et al. (1988).

Transformant Selection

Selection was started after 3–5 days post-bombardment on solid medium containing 3–4 mM glyphosate, or 1mg/l methotrexate or 3 mg/l bialaphos. Tissue was subcultured onto fresh selection medium every 14 days. After 6–8 weeks of selection, resistant colonies start appearing. Later, individual colonies were transferred to fresh media containing 4 mM glyphosate, or 1mg/l methotrexate or 3 mg/l bialaphos for further growth and analysis.

DNA Isolation and PCR Analysis

DNA was prepared from callus tissues according to K. Cone (Maize Genetics Cooperation Newsletter, 68:1989). CP4 and GOX PCR reactions were run using a kit from Perkin Elmer Cetus. Amplification parameters were 95°C denaturing for 1 min, 50°C annealing for 2 min, and 72°C extension for 3 min for 30 cycles.

RESULTS AND DISCUSSION

Cell Culture and Genotype Evaluation

Bobwhite is the best genotype for the rapid production of embryogenic callus from immature embryos from 10 diverse genotypes screened. After 4 weeks in culture Bobwhite genotype produced a significantly greater amount of primary embryogenic callus(PEC) on the MS 2,4-D medium than the other genotypes. This genotype produced between 1.6 to 4 times as much callus as the immature embryos from the next most responsive genotype

(UC702). Bobwhite produced 7 times more PEC than Pavon (control genotype) on the MS 2,4-D medium.

Transient Expression Assay

Intact cells of *Triticum aestivum* cv. Mustang bombarded with equimolar amounts of two plasmid : pMON19574 encoding CP4/GOX or pMON19335 encoding DHFR or pMON19477 encoding the BAR gene and BC17 encoding the anthocyanin genes, showed expression of these genes as determined visually by the accumulation of purple-red pigments into intact cells. After 48 hr of incubation in the dark, isolated purple cells could be visualized, but more generally 3–6 adjacent cells developed pigmentation, with a central cell being the most pigmented. On average more than 3000 purple-red spots were observed from 0.5 ml PCV of suspension cells. Similar to suspension cells, the number of purple-red units per immature embryo varies from 80–100 and per embryogenic calli from 100–250 units. The control samples bombarded by particles without DNA did not show any pigment in the cells.

Sensitivity of Selective Agents to Wheat Cells for Selection of Transformants

To be able to judge which of these selection regimes is most efficient for wheat the sensitivity of the Mustang suspension cells to the selective agents was determined. On medium containing 3 mM glyphosate, growth of suspension cells was inhibited 90%, whereas a 4 mM glyphosate concentration resulted in total cessation of growth followed by death of cells within 10–14 days. Similarly, bialaphos 3mg/l and methotrexate 1mg/l were needed to stop growth 90–95%(as measured visually) within 10–14 days.

Glyphosate Selection of Stably Transformed Callus

Five days after bombardment, selection was started on medium containing 3–4mM glyphosate. Colonies resistant to glyphosate continued to grow and reached 100–300μm in size after 4–5 weeks of selection, whereas the non-transformed colonies ceased growth. After approximately eight weeks of selection the average number of resistant calli was approximately 30 under 3–4 mM glyphosate selection, of which approximately 16 were purple-red (anthocyanin positive) giving 53% co-expression frequency. The pattern of cells accumulating, anthocyanin ranged from virtually all purple-red calli to cells with patchy expression of purple-red.

Expression of Anthocyanin Genes with Different Selectable Markers

Similar to glyphosate selection, suspension cells bombarded with BC17 and plasmid containing DHFR or BAR showed the same number of transient spots as observed with the glyphosate vector (approximately > 3000 units). Similar, to glyphosate selection, anthocyanin producing calli can be visually selected after 4–5 weeks, with each selection agent. After 6–8 weeks of selection resistant colonies grew fast and reached the 2–4 mm size. From 3–4 independent bombardment filters, the average number of methotrexate and bialaphos resistant calli was approximately 25 of which 10–12 were anthocyanin positive, giving a 40–45% co-expression of unlinked genes.

PCR Analysis

Confirmation of the presence of the CP4 and GOX genes in the glyphosate resistant and anthocyanin accumulating calli was done by DNA amplified by PCR analysis. Amplification was with primers for a segment of the CP4 and GOX which is present in pMON19574. All six calli analyzed were positive for CP4 (360 bp) and GOX (280 bp).

Expression of Anthocyanin Genes in Embryogenic Callus

Bobwhite immature embryo derived embryogenic calli showing anthocyanin accumulation after 48 hours were allowed to grow. After 7–14 days post-bombardment embryogenic calli expressing anthocyanin were transferred onto selection medium. Under selection, anthocyanin expressing cells proliferated and underwent repeated division. Isolated purple-red calli were recovered under selective pressure. The callus lines exhibited a range of color phenotypes from solid dark-red to expression only in a specific portion of the embryogenic callus. Embryogenic calli expressing anthocyanin could be selected visibly and transferred onto regeneration medium. Occasionally, regenerating shoots expressing anthocyanin (purple-red) were observed. Plantlets were regenerated from these events. The phenotype of the leaves expressing anthocyanin varied from solid red to striped.

CONCLUSION

This study provides unequivocal proof demonstrating stable transformation of wheat. The proof is anthocyanin accumulation, selection onto glyphosate, bialaphos and methotrexate; enzyme assay and PCR analysis for CP4/GOX following microprojectile bombardment of suspension cells. The procedure described yields approximately 30 transformants per bombardment with glyphosate selection. Similarly, using DHFR and BAR genes with methotrexate and bialaphos as selective agents respectively, an average of 25 independent resistant callus lines were obtained.

In the regenerable system, of ten genotypes tested Bobwhite (*Triticum aestivum*) cultivar produces a highly embryogenic callus from immature embryos. After bombardment of young callus from Bobwhite immature embryos with DNA containing the BC17 genes we have demonstrated both transient and stable expression of the anthocyanin genes in wheat callus and plant tissue.

REFERENCES

Benfey, P. Takatsuji, H. Ren, L. et al. (1990). Sequence requirements of the 5-enolpyruvylshikimate-3-phosphate synthase 5′-upstream region for tissue-specific expression in flowers and seedlings. *Plant Cell.* 2:849–856.

Bowen, B. (1992). Anthocyanin genes as visual markers in transformed maize tissue. GUS Protocols: Using the GUS as a Reporter of Gene Expression.:163–177.

Callis, J. Fromm, M. Walbot, V. (1987) Introns increase gene expression in cultured maize cells. 1: 1183–1200.

Dunder, E. Dawson, J. Brewer, J. Genes Dev. et al. (1993). Transgenic anthocyanin color phenotypes produced in callus, plants and progeny of maize. 35th Annual Maize Genetic Conf. pp.15.

Hu, C. Chee, P. Chesney, R. et al. (1990). Intrinsic GUS-like activities in seed plants. *Plant Cell* Rep 9:1–5.

Klein, T. Gradziel, T. Fromm, M. et al. (1988). Factors influencing gene delivery into *Zea mays* cells by high-velocity microprojectibles. *Bio/Technol.* 6:559–563.

Ludwig, S. Bowen, B. Beach, L. et al. (1990). A regulatory gene as a novel visible marker for maize transformation. *Science* 247:449–450.

Murashige, T. Skoog, F. (1962). A revised medium for rapid growth and bioassays with tobacco tissue culture. *Physiol. Plant* 15:473–497.

Redway, F. Vasil, V. Lu, D. Vasil, I. (1990).Characterization and regeneration of wheat (*Triticum aestivum* L.) embryogenic cell suspension cultures.*Plant Cell Rep.* 8:714–717.

Tor, M. Mantell, S. Ainsworth, C. (1992).Endophytic bacteria expressing β-glucuronidase to cause false positives in transformation of *dioscorea species. Plant Cell Rep* 11:452–456.

Wong, J. Walker, L. Klein, T. (1991). Anthocyanin biosynthesis can be activated in Diverse Monocot Species by Regulatory Genes from Maize. *33rd Annual Maize Genetic Conf.* pp. 68.

SECTION II

GENETIC ENGINEERING OF CEREAL PROTEIN QUALITY

IMPROVEMENT OF BARLEY AND WHEAT QUALITY BY GENETIC ENGINEERING

P. R. Shewry,[1] A. S. Tatham,[1] N. G. Halford,[1] J. Davies,[2] N. Harris,[2] and M. Kreis[3]

[1]Department of Agricultural Sciences
AFRC Institute of Arable Crops Research
University of Bristol
Long Ashton Research Station
Long Ashton, Bristol, BS18 9AF, United Kingdom

[2]Durham University
Department of Botany
University of Durham
Science Laboratories
South Road, Durham DH1 3LE, United Kingdom

[3]Université de Paris-Sud
Biologie du Developpement des Plantes
Batiment 430
F-91400 Orsay Cedex, France

INTRODUCTION

Although the mature cereal grain consists predominately of starch with only about 10–15% of protein, it is the protein fraction which is largely responsible for quality. In the case of wheat the quality for breadmaking, for other baked goods (biscuits, wafers, cakes and pastries) and for noodles and pasta is determined by the gluten protein fraction, while the nutritional quality of barley and wheat for monogastric livestock is limited by the low contents of essential amino acids in the prolamin storage proteins. In the case of malting barley the quality is determined not only by the proteins that accumulate during grain development but also by enzymes synthesised *de novo* during germination. In the present paper we will briefly review our studies of grain quality in barley and wheat, and speculate how quality parameters can be manipulated by genetic engineering.

Improvement of Cereal Quality by Genetic Engineering, Edited by
Robert J. Henry and John A. Ronalds, Plenum Press, New York, 1994

FEED QUALITY OF BARLEY

The quality of barley for animal feed is limited by the low contents of lysine and threonine present in the prolamin storage proteins (hordeins). These proteins account for about half of the total grain proteins, resulting in levels of lysine and threonine in the mature grain of about 3.1 and 3.3 g/100g of protein respectively compared with WHO recommended levels of 5.5 and 4.0 g/100g (Ewart, 1967; FAO, 1973). Because lysine is the first limiting amino acid this has received most attention.

Early studies of barley quality focused on two areas: basic studies of the biochemistry, genetics and molecular biology of the hordein storage proteins, and empirical attempts to improve quality by the identification and incorporation of mutant "high lysine" genes. Neither of these approaches has led to the successful production of high lysine cultivars. However, analysis of one mutant high lysine line (Hiproly) has led to the identification of high lysine proteins, the amounts of which could possibly be increased by genetic engineering.

In particular, two inhibitors of chymotrypsin have been isolated which have M_rs of about 8,800 (CI-1) and 9,400 (CI-2) and contain about 9.5 and 11.5 mol % lysine respectively (Hejgaard and Boisen, 1980). Their combined activities are increased from about 0.7 units/g in normal lines to between 4.5 and 5.6 units/g in Hiproly and derived lines (Hejgaard and Boisen, 1980), without any obvious effects on grain development or digestibility.

Several cDNAs for CI-1 and CI-2 have been isolated (Williamson *et al*, 1987; 1988) (Fig.1), and a single gene for CI-2 (Peterson *et al*, 1991). Expression of chimaeric CI-2/*GUS* constructs in transgenic tobacco plants established that about 1100 bp of 5′ upstream

Amino Acid Sequences of
CI-1 and CI-2 of Barley

```
CI - 1A   M S S M E G S V L K Y P E P T E G S I G
     1B   M R S M E G S V P K Y P E P T E G S I G
CI - 2A   M S S V E - - - - K K P E G V N T G A G
     2B                                 * D C L C

CI - 1A   A S S A - K T S W P E V V G M S A E K A
     1B   A S G A - K R S W P E V V G M S A E K A
CI - 2A   D R H N L K T E W P E L V G K S V E E A
     2B   D C Q N Q K T E W P E L V E K S V E E A

CI - 1A   K E I I L R D K P N A Q V E V I P V D A
     1B   K E I I L R D K P D A Q I E V I P V D A
CI - 2A   K K V I L Q D K P E A Q I I V L P V G T
     2B   K K V I L Q D K P E A Q I I V L P V G T

CI - 1A   M V H L N F D P N R V F V L V - - - - A
     1B   M V P L D F N P N R I F I L V - - - - A
CI - 2A   I V T M E Y R I D R V R L F V D K L D N
     2B   I V T M E Y R I D R V R L F V D R L D N

CI - 1A   V A R T P T V G
     1B   V A R T P T V G
CI - 2A   I A Q V P R V G
     2B   I A Q V P R V G
```

* sequence incomplete

Figure 1. The amino acid sequences of barley chymotrypsin inhibitors CI-1 and CI-2 deduced from the nucleotide sequences of cDNAs (Williamson *et al*, 1987; 1988). Two subfamilies of cDNAs for each protein were isolated, called A and B. The boxes indicate identical residues in all four proteins.

sequence was sufficient to confer endosperm-specific expression. However, it may be preferable to use a stronger endosperm-specific promoter, for example an HMW subunit gene promoter from wheat (see below), if the CI-2 coding region is to be used to increase the lysine content of transgenic cereal seeds.

MALTING QUALITY OF BARLEY

The malting quality of barley is determined by factors operating during grain development and during the conversion of grain to malt. The latter will be discussed in a later chapter in this volume (Fincher), but I would like to briefly mention two aspects of grain development. The first is the activity of β-amylase. Whereas most hydrolytic enzymes (including α-amylase) are synthesised *de novo* during grain germination, β-amylase is synthesised during grain development and stored in the mature grain. There is, to the best of my knowledge, no evidence to suggest whether β-amylase is present in excess during germination and malting or is limiting. However, the availability of a cloned cDNA (Kreis *et al*, 1987) and gene (unpublished results of M. Kreis and M. Thomas) will allow this to be explored in transgenic barley plants. In addition it will also be possible to determine whether there is any effect when the protein is expressed during grain germination (controlled, for example, by an α-amylase or β-glucanase promoter) in addition to or instead of during grain development.

The second aspect of grain development is the role of the hordein storage proteins. Polymeric hordeins are known to be the major components of "gel protein," the amount of which is inversely correlated with malting quality (Smith and Lister, 1983). However, the precise genetic basis for this correlation, and its relationship to allelic variation at the hordein structural loci, are not known. In addition, Millet *et al*, (1991) have presented preliminary evidence that differences in malting quality may be associated not only with the amounts of polymeric hordeins, but also with their spatial distribution within the grain. Thus the peripheral layers of the starchy endosperm contained a higher proportion of polymeric hordeins in a poor quality cultivar than in a good quality cultivar. We are currently investigating this phenomenon in more detail using a combination of immunocytochemical analyses of serial sections of developing endosperms (Fig. 2) (see Davies *et al*, 1993) and biochemical analyses of pearling fractions from mature seeds. The possibility that spatial aspects of hordein deposition may affect malting quality means that it is essential to understand the fine control of hordein gene expression in the developing endosperm in addition to defining sequences which confer broad endosperm specificity (Marris *et al*, 1988).

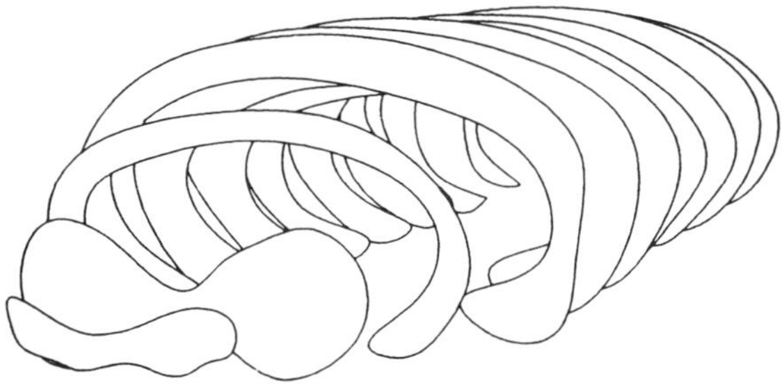

Figure 2. Reconstruction of the immunolocalisation of B hordein in a 7 day old developing endosperm of barley, based on serial sections. Taken from Davies *et al* (1993).

BREADMAKING QUALITY OF WHEAT

The breadmaking quality of wheat is determined largely by the amount and properties of the gluten proteins. These correspond to the prolamin storage proteins, and account for about half of the total grain proteins.

Gluten forms a network in the dough and has a unique combination of elasticity and viscosity. A precise balance of these properties is crucial for breadmaking, and poor quality wheats usually contain gluten which is insufficiently elastic. Gluten consists of over 50 individual proteins which are usually classified into two groups. The gliadins are single monomeric proteins which associate by strong non-covalent forces, and are largely responsible for gluten viscosity. The glutenins consist of individual polypeptides which are assembled into high M_r polymers by inter-chain disulphide bonds, and are associated with gluten elasticity. Both the gliadin and glutenin proteins can be further sub-divided into groups: the gliadins into α, γ and ω-types, and the glutenin subunits into high molecular weight (HMW) and low molecular weight (LMW) groups.

Although all the individual gluten proteins contribute to its structure and functionality, attention has focused on one specific group: the HMW subunits of glutenin. This is because variation in breadmaking quality is correlated with allelic variation in the HMW subunit composition (see Payne, 1987). In addition the HMW subunits are only present in high M_r polymers, the amounts of which are also correlated with breadmaking quality (Field *et al*, 1983).

Genetic studies have shown that the HMW subunits are encoded by loci on the long arms of the group 1 chromosomes (1A, 1B and 1D), each locus consisting of two genes encoding one high M_r x-type and one low M_r y-type subunits (Payne, 1987). However, not all of these genes are expressed. Thus 1Ay subunit genes are never expressed in hexaploid bread wheats, while 1Ax and 1By subunit genes are only expressed in some cultivars. As a result the number of HMW subunits varies from three to five, compared with a theoretical maximum of six (Fig. 3). (Harberd *et al*, 1986). In addition all the expressed subunits occur in allelic variants which differ in their mobilities on SDS-PAGE and are given numbers (Payne and Lawrence, 1983).

Genomic clones have been isolated for a total of nine different HMW subunits, including allelic forms associated with good and poor quality (see Anderson *et al*, 1988; Shewry *et al*, 1989; 1992). They encode proteins consisting of between 481 and 696 residues, with M_rs of 67,500 to 88,100. The proteins all have similar structures, with central repetitive domains flanked by shorter non-repetitive domains at the N-terminus (81 to 104 residues) and C-terminus (42 residues in all subunits). The repetitive domains consist of random and interspersed repeats based on hexapeptide, nonapeptide and, in y-type subunits only, tripeptide repeat motifs (Fig. 4).). It is notable that most, and in some cases all, of the cysteine residues are located in the N- and C-terminal domains. The repeated sequences appear to form an unusual spiral supersecondary structure, based on repeated β-turns. This has been visualised directly by scanning tunnelling microscopy (Miles *et al*, 1991) (Fig. 5), and results in an extended rod-like structure for the whole molecule.

The HMW subunits appear to have quantitative and qualitative effects on breadmaking quality. Good quality is associated with the presence of a 1Ax subunit, as opposed to the silent or null allele (Payne, 1987). This is associated with an increase in total HMW subunit protein, from about 8% of the total extracted protein in cultivars with 1Bx, 1By, 1Dx and 1Dy subunits to about 10% in cultivars which also contain a 1Ax subunit (Halford *et al*, 1992). It is probable that increased total HMW subunit protein results in a higher proportion of elastic high M_r polymers. However, variation in quality is also associated with allelic vari-

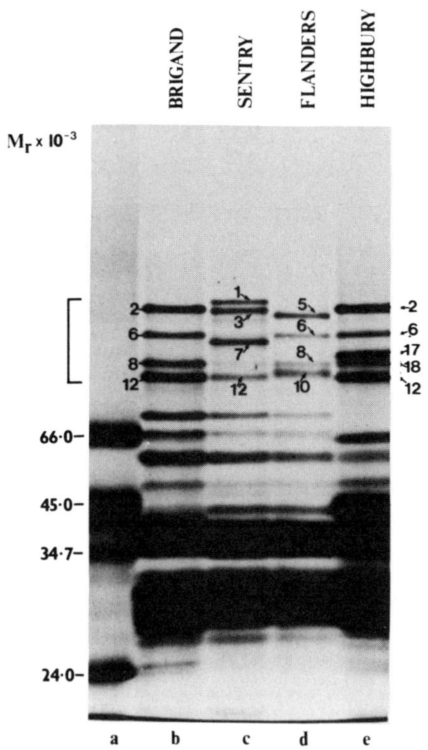

Figure 3. SDS-PAGE of total prolamin fractions from four cultivars of wheat, illustrating polymorphism in HMW subunits of glutenin. The subunits present are 1Ax1, 1Bx6, 1Bx7, 1Bx17, 1Dx2, 1Dx5, 1By8, 1By18, 1Dy10 and 1Dy12.

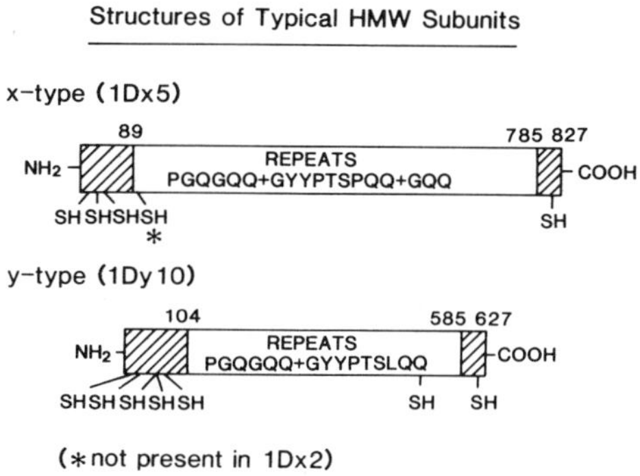

Figure 4. Schematic structures of the "good quality" HMW subunits 1Dx5 and 1Dy10. The positions of cysteine residues are indicated by SH. The asterisk indicates a cysteine residue present in 1Dx5 but not 1Dx2. Subunit 1Dy12 has a similar content of cysteine residues to 1Dy10.

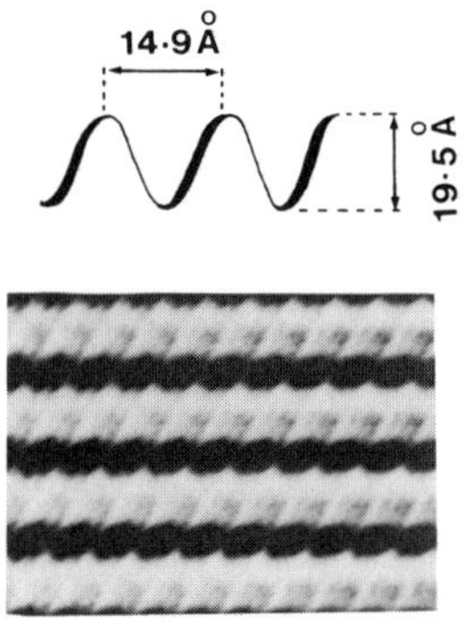

Figure 5. STM image (below) of HMW subunit 1Bx20, showing aligned rods with diagonal striations. The diameter of the rods is about 19.5 Å and the pitch of the spiral 14.9 Å, as shown in the diagram above. Taken from Shewry *et al*, 1992.

ation in expressed HMW subunits; for example with subunits 1Dx5+1Dy10 as opposed to subunits 1Dx2+1Dy12. In this case the difference is qualitative, and may be related to the presence of an additional cysteine residue in subunit 1Dx5 compared to subunit 1Dx2 (see Greene *et al*, 1988; Shewry *et al*, 1992). This could lead to the presence of more highly cross-linked, and hence more elastic, polymers.

These results indicate that it should be possible to improve the quality of wheat by genetic engineering, by inserting additional genes for HMW subunits. Since each gene accounts on average for about 2% of the total protein only one or two additional copies should be sufficient to result in a significant improvement in quality. In addition the transfer of mutated forms of HMW subunit genes should allow the fine-tuning of gluten structure for specific end uses.

CONCLUSIONS

Studies carried out over the past few years have provided a detailed understanding of some aspects of grain quality (e.g., for feed and breadmaking), but further work is required to establish the molecular basis for other quality traits (e.g., malting quality). In the former case the recent development of efficient transformation systems for small grain cereals (Vasil *et al*, 1992; Weeks *et al*, 1993; Barcelo *et al*, 1993) should enable improvements in quality to be made in the near future. In the latter case gene transfer technology will allow the quality traits to be explored in detail by determining the consequences of the up or down regulation of specific genes. The next few years should prove to be very exciting for cereal biotechnologists.

REFERENCES

Anderson, O.D., Halford, N.G., Forde, J., Yip, R., Shewry, P.R. and Greene, F.C. (1988) Structure and analysis of the high molecular weight glutenin genes from *Triticum aestivum* L. cv. Cheyenne. *Proceedings of the 7th International Wheat Genetics Symposium* IPSR, Cambridge, U.K. pp. 699–704.

Barcelo, P., Hazel, C., Becker, D., Martin, A. and Loerz, H. (1993) Fertile transgenic cereal plants obtained by particle bombardment of inflorescence tissue. Submitted.

Davies, J.T., Shewry, P.R. and Harris, N. (1993) Spatial and temporal patterns of B hordein synthesis in developing barley (*Hordeum vulgare* L.) caryopses. *Cell Biology International Reports* **17**, 195–203.

Ewart, J.A.D. (1967) Amino acid analyses of cereal flour proteins. *Journal of the Science of Food and Agriculture* 18, 548–552.

FAO. Energy and Protein requirements, *FAO Nutritional Meet. Rep. Ser. No.52, WHO Technical Report Ser. No. 522*, Rome, 1973.

Field, J.M., Shewry, P.R. and Miflin, B.J. (1983) Solubilization and characterization of wheat gluten proteins; correlations between the amount of aggregated proteins and baking quality. *Journal of the Science of Food and Agriculture* 34, 370–377.

Greene, F.C., Anderson, O.D., Yip, R.E., Halford, N.G., Malpica-Romero, J-M. and Shewry, P.R. (1988) Analysis of possible quality-related sequence variations in the 1D glutenin high molecular weight subunit genes of wheat. *Proceedings of the 7th International Wheat Genetics Symposium* IPSR, Cambridge, U.K. pp.735–740.

Halford, N.G., Field, J.M., Blair, H., Urwin, P., Moore, K., Robert, L., Thompson, R., Flavell, R.B., Tatham, A.S. and Shewry, P.R. (1992) Analysis of HMW glutenin subunits encoded by chromosome 1A of bread wheat (*Triticum aestivum* L.) indicates quantitative effects on grain quality. *Theoretical and Applied Genetics* 83, 373–378.

Harberd, N.P., Bartels, D. and Thompson, R.D. (1986) DNA restriction fragments variation in the gene family encoding high molecular weight (HMW) glutenin subunits of wheat. *Biochemical Genetics* 24, 579–596.

Hejgaard, J. and Boisen, S. (1980) High lysine proteins in Hiproly barley breeding: identification, nutritional significance and new screening methods. *Hereditas* 93, 311–20.

Kreis, M., Williamson, M., Buxton, B., Pywell, J., Hejgaard, J. and Svendsen, I. (1987) Primary structure and differential expression of β–amylase in normal and mutant barleys. *European Journal of Biochemistry* 169, 517–25.

Marris, C., Gallois, P., Copley, J. and Kreis, M. (1988) The 5′-flanking region of a barley B hordein gene controls tissue and development specific CAT expression in tobacco plants. *Plant Molecular Biology* 10, 359–366.

Miles, M.J., Carr, H.J., McMaster, T., Belton, P.S., Morris, V.J., Field, J.M., Shewry, P.R. and Tatham, A.S. (1991) Scanning tunnelling microscopy of a wheat gluten protein reveals details of a spiral supersecondary structure. *Proceedings of the National Academy of Sciences USA* 88, 68–71.

Millet, M.-O., Montembault, A. and Autran, J.-C. (1991) Hordein compositional differences in various anatomical regions of the kernel between two different barley types. *Sciences des Alimentations* 11, 155–161.

Payne, P.I. and Lawrence, G.D. (1983) Catalogue of alleles for the complex gene loci *Glu-A1, Glu-B1* and *Glu-D1* which code for high molecular weight subunits of glutenin in hexaploid wheat. *Cereal Research Communications* 11, 29–35.

Payne, P.I. (1987) Genetics of wheat storage proteins and the effect of allelic variation on breadmaking quality. *Annual Review of Plant Physiology* 38, 141–153.

Peterson, D.M., Forde, J., Williamson, M.S., Rhode, W. and Kreis, M. (1991) The nucleotide sequence of a chymotrypsin inhibitor-2 gene of barley (*Hordeum vulgare*). *Plant Physiology* 96, 1389–90.

Shewry, P.R., Halford, N.G. and Tatham, A.S. (1989) The high molecular weight subunits of wheat, barley and rye: genetics, molecular biology, chemistry and role in wheat gluten structure and functionality. In *Oxford Surveys of Plant Molecular and Cell Biology* 6 (B.J. Miflin, Ed.) pp. 163–219. Oxford University Press, Oxford.

Shewry, P.R., Halford, N.G. and Tatham, A.S. (1992) The high molecular weight subunits of wheat glutenin. *Journal of Cereal Science* 15, 105–120.

Smith, D.B. and Lister, P.R. (1983) Gel forming proteins in barley grain and their relationship with malting quality. *Journal of Cereal Science* 1, 229–239.

Vasil, V., Castillo, A.M., Fromm, M.E. and Vasil, I.K. (1992) Herbicide resistant fertile transgenic wheat plants obtained by microprojectile bombardment of regenerable embryogenic callus. *Bio/Technology* **10**, 667–675.

Weeks, J.T., Anderson, O.D. and Blechl, A.E., (1993) Rapid production of multiple independent lines of fertile transgenic wheat. *Plant Physiology*. 102, 1077–1089.

Williamson, M.S., Forde, J., Buxton, B. and Kreis, M. (1987) Nucleotide sequence of barley chymotrypsin inhibitor-2 (C1-2) and its expression in normal and high-lysine barley. *European Journal of Biochemistry* 165, 99–106.

Williamson, M.S., Forde, J. and Kreis, M. (1988) Molecular cloning of two isoinhibitor forms of chymotrypsin inhibitor 1 (C1-1) from barley endosperm and their expression in normal and mutant barleys. *Plant Molecular Biology* 10, 521–35.

PROGRESS TOWARDS GENETIC ENGINEERING OF WHEAT WITH IMPROVED QUALITY

Olin D. Anderson,[1] Ann E. Blechl,[1] Frank C. Greene,[2] and J. Troy Weeks[1]

[1]U. S. Department of Agriculture
Agricultural Research Service
Western Regional Research Center
800 Buchanan Street
Albany, California, 94710

[2]U. S. Department of Agriculture
Agricultural Research Service
Russell Agricultural Research Center
P.O. Box 5677
College Station Road
Athens, Georgia 30613

SUMMARY

Progress is reported on the goal of our laboratory to use biotechnological approaches to improve wheat quality. The wheat gene family we have chosen for our initial efforts is that of the high-molecular-weight (HMW) glutenins. These polypeptides play a critical role in the formation of the disulfide cross-linked protein matrix correlated with dough functionality. The development of sufficiently reliable wheat transformation methodology will allow direct testing of the effect of native and modified HMW-glutenin genes on wheat quality. Our laboratory now has the capability of transforming wheat reliably enough to carry out such studies. We have used the highly regenerative hard white spring cultivar Bobwhite. DNA is delivered *via* bombarding immature embryos with gold particles coated with the *bar* gene (which confers resistance to the herbicides bialaphos and Basta). Over a nine month period we produced 9 lines of transformed wheat. The criteria for successful transformation was herbicide resistance, transgene encoded enzyme activity, integration of transforming DNA into high-molecular-weight wheat DNA, and co-segregation of the DNA and phenotype in the T_1 and T_2 generations. Wheat transformation is a key component in a schema proposing a complete set of approaches to apply biotechnology to questions of wheat quality *via* manipulation of the HMW-glutenin genes. A second important component is the heterologous expression of HMW-glutenin genes, and the use of the resulting proteins in the study of glutenin physical chemistry and dough supplementation experiments. While the techniques

Improvement of Cereal Quality by Genetic Engineering, Edited by
Robert J. Henry and John A. Ronalds, Plenum Press, New York, 1994

are still in the process of fine-tuning, we believe all the necessary tools are now available to attempt modification of the physical properties of dough.

INTRODUCTION

Wheat cultivation began somewhere in the middle east and has been ongoing for perhaps as long as 10,000 years. Today, wheat, rice and maize form the foundation of the human diet in every corner of the world. Of these three, wheat is the most widely grown, but since multiple rice crops are possible in much of rice's tropical planting range, wheat and rice vie yearly for the title of the world's crop of the highest tonnage. However, since wheat has a significantly higher protein content than rice, wheat is the single greatest source of dietary protein in the human diet. The importance of wheat and the other cereals, both because they are favorites for human consumption and because of their economic value, have led to their prominence in visions of the future of biotechnology. The full impact of biotechnology on agriculture will occur only when all the necessary technologies are available to cereal scientists, a prerequisite that is yet to be fulfilled. There are many aspects to biotechnological approaches to wheat modification, but the technical requirements can be divided into two basic categories. The first is transformation; i.e., there must be protocols to introduce exogenous genes directly into the wheat genome. Once this DNA is integrated into the wheat genome, it must be stable and passed on to the subsequent generations. Without this technology, the full power of molecular biology be cannot be applied to this critical crop. The result would be that the development wheat science and crop utilization would languish compared to other plant systems which can be transformed. Moreover, it is not even sufficient simply that wheat is transformable. The protocols must be efficient and accessible to many laboratories. In spite of wheat's importance, research and development must balance the significance of the crop under study with the technical ease of manipulation. A procedure which is excessively research intensive, poorly reproducible, and too dependent on individual researcher expertise will not permit the maximum application of modern molecular techniques to wheat bioengineering.

The second requirement for bioengineering wheat is the possession of genes that researchers have a reasonable expectation will confer positive traits if these genes can be transformed into wheat. Eventually it will be possible to alter the biochemical pathways of wheat (both modification of existing pathways and introduction of foreign or chimeric pathways), but initial transformations will concentrate on traits which can be changed with the addition of single or a small number of genes. Such genes will include those affecting herbicide resistance, insect and viral resistance, storage proteins, and starch metabolism.

WHEAT TRANSFORMATION

A definition of transformation that emphasizes its goal is: the laboratory transfer of genes directly into a recipient genome, functional integration into that genome such that the appropriate developmental expression occurs, and stable transmission of the gene and its regulatory characteristics to subsequent generations. When such transformation is possible, attempts to genetically engineering the organism are possible. The monocotyledonous plants have proven to be more difficult to successfully transform than the dicots. The basis of this difficulty is not known, but the unfortunate result has been that the most important crops have not fully participated in advances in plant genetics and molecular biology, and the application to new crop development has been pushed further and further into the future. The first breakthrough in the block to monocot transformation occurred with the report by Toriyama et al. (1988) of the first successful rice transformation. This report was quickly followed by other rice successes (Shimamoto et al., 1989; Zhang and Wu, 1989). These results were followed by transformation reports for maize (Fromm et al., 1990; Gordon-Kamm et al., 1990) and oats (Somers et al., 1992). Wheat finally was shown to be transformable with the report of a single transgenic plant line (Vasil et al., 1992). Once the principle thatwheat transformation was possible was established, it became necessary to improve the methodology. It has been goal of our laboratory to develop wheat transformation protocols that would allow any suitably equipped laboratory to undertake a research program involving wheat transformation. To reach this goal, the protocols needed to be efficient enough to yield multiple transformed wheat lines without excessive effort; be reproducible on a regular basis; yield fertile lines which pass on the genotypes and phenotypes to successive generations, require no special tissue culture expertise outside of basic principles; and require no material or equipment not publically available. We have recently reported a rapid and efficient method of producing transgenic wheat lines that meets all of the above criteria (Weeks et al., 1993). The transformation and regeneration protocol is summarized in Fig. 1. Immature embryos are excised approximately 15 days after anthesis and cultured, scutellum side up, on media for five days. At this point cell proliferation has increased the tissue mass, particularly at the edges of the scutellum. The tissue is then bombarded with 0.1 µm gold particles coated with the genes to be inserted into the wheat genome (initial experiments used the *bar* gene to confer phosphinothricin resistance, and the *uidA* gene for the colorimetric marker enzyme β-glucuronidase). Immediately after bombardment the tissue is placed on 1 mg/L of the herbicide bialaphos (from Meiji-Seika). Over the next several months the tissue is transferred to fresh media every two weeks and observed for shoot formation. Tissues developing shoots are dissected away from the main tissue mass and placed on a rooting medium. This is the point in the transformation protocol that thus far has proved to be diagnostic of stable integration of the incoming genes. If a complex root network forms within two weeks, transformation has been successful, as confirmed by later analyses. After rooting the plantlets are transferred to soil and the greenhouse to complete development and seed set. There are two intervals noted in Figure 1. The first is the time, approximately six months, to complete a life cycle from the original anthesis to anthesis of the T_0 plants. The second interval is the time from embryo excision to the positive identification of transformed plants, and is on average 4 months (although it has been as short as 2.5 months). To gauge the relative efficacy of the protocol, we estimate that a researcher with basic tissue culture experience, devoting full time to wheat transformation, and with technical assistance, can produce 30–50 independent lines of transformed wheat in one year. More time and/or resources would be necessary to analyze the resulting transgenic lines. Where extensive analysis is necessary, the practical number of lines produced would be much less.

We believe it is now feasible to initiate wheat engineering experiments. Not that all problems are solved. While transformation is now feasible, there are a number of technical aspects needing improvement and questions yet to be answered.

1. Efficiency. Our current protocol is adequate if a single gene of significant importance is to be inserted into the wheat genome. However, it is not efficient enough if very large numbers of transformants are needed (unless large resources are available). Therefore, the efficiency needs at least a ten-fold improvement, a goal we believe is within reach.

2. Position Effects. Our results confirm what has been seen in other animal and plant systems: i.e., the expression levels have no correlation to the gene copy number. The hypoth-

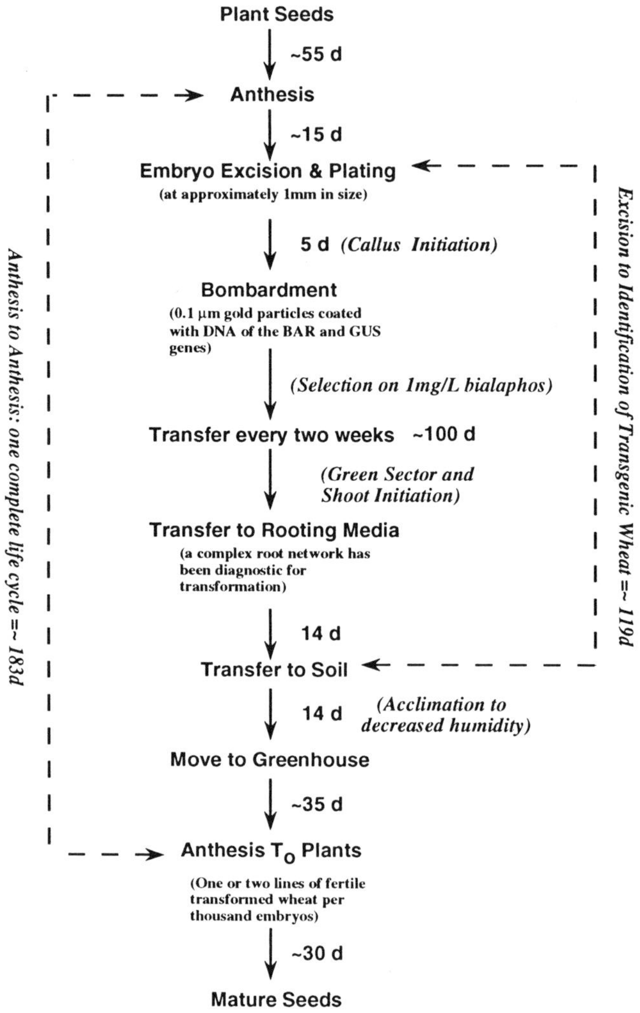

Figure 1. Summary of wheat transformation protocol. A summary flowchart of the wheat transformation protocol used in our laboratory is presented along with notes on critical steps. Dotted lines indicate two important time intervals: i.e., the time for one complete life-cycle and the time from experiment initiation until identification of transformed plantlets.

esis is that when multiple copies insert randomly into the genome, the site of insertion limits the expression potential of the transgenes. This "position effect" complicates engineering efforts, particularly when the goal is high levels of expression such as in the seed. The problem can be handled in three ways: screen enough transgenic lines to find the desired levels of transgene expression; include DNA elements in the transforming construct to eliminate the position effect (such as nuclear matrix-attachment elements reported to eliminate position effects in animal systems; McKnight et al., 1992); target the insertion to regions of the genome competent for high levels of gene expression.

3. Inactivation of the Transgene. It is known that a successfully transferred gene can become inactivated by methylation (Meyer et al., 1993) or be inactive due to the possibly related phenomenon of co-suppression (Hart et al., 1992). Neither of these mechanisms of inactivation is well understood and there is no method of predicting their occurrence for any specific gene or transgenic line of plants. Wheat transformation is too new to have experience with inactivation problems, and we must simply deal with such events if and when they occur.

4. Genotype Specificity. For maximum efficiency in developing new cultivars we need to use either current high quality cultivars or at least elite breeding lines. Thus the wheat breeder could introduce new genes quickly. If the transformed cultivar requires extensive backcrossing with elite lines, the additional years required would make some transfers impractical. Our protocol should be successful with any wheat type or cultivar which can be regenerated from immature embryo callus cultures. It is known that different cultivars have a range of response to tissue culture (Sears and Deckard, 1982; J. Driver, A. Guenzi, and T. Peeper, unpublished) and the less responsive may not be useable with the current methodology. These less-responsive cultivars would be within reach of transformation technology if the efficiency of the method were improved. Genotype specificity does not preclude wheat bioengineering, but a solution will speed development of new varieties.

The Wheat High-Molecular-Weight Glutenin Genes as Targets for Genetic Engineering

The results presented at this meeting, from both our laboratory and other laboratories, show that wheat transformation technology is now a reality. While many technical questions remain, a now more central issue is what genes to use in attempts to improve wheat quality and utilization.

Wheat is unique because the mixture of flour and water produces a dough with unusual physical properties such as visco-elasticity. The rheological properties of dough can be attributed to many flour components, but the clearest association is with the storage proteins of the grain and the flour. These proteins are divided into several families, one of which, the high-molecular-weight (HMW) glutenins, correlates the best with wheat quality (Payne et al., 1981; Payne 1987). The molecular basis of the contribution of these proteins to dough visco-elasticity is not well defined, but molecular biology has begun to provide information that will eventually lead to a better understanding (reviewed by Shewry et al., 1991). Our laboratory and collaborators have isolated and sequenced a complete set of HMW-glutenin genes from a singlecultivar, the hard red winter wheat Cheyenne (Anderson et al., 1988), including the two genes with the highest correlation to good wheat quality (Anderson et al., 1989). From the deduced amino acid sequences of these genes, a model of HMW-glutenin subunit organization is diagrammed in Figure 2. The dominant feature is the central repetitive domain composed of 60–90 repeats of two or three simple peptide motifs. Computer predictions of secondary structure suggests a series of β turns. It has been theorized (Shewry et al., 1992) that a β-spiral structure results which may relate to dough visco-elasticity similar to what has been proposed for the contribution of a β-spiral to the organization of the vertebrate connective tissue protein elastin (Urry, 1984). A model of dough visco-elasticity could include elastic properties contributed by the repetitive domains to a matrix of glutenin subunits. As simple as this model may be, there is as yet no direct evidence of its validity.

The second important feature of these proteins is the terminal placement of the cysteine residues. It is known that fundamental to dough functionality is the integrity of a highly

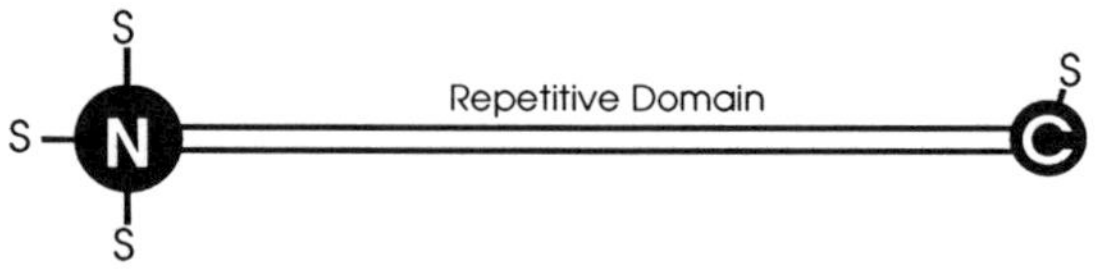

Figure 2. Diagrammatic representation of a wheat high-molecular-weight glutenin subunit. These subunits form a critical role in the disulfide cross-linked protein network necessary for wheat quality characteristics. The two main features of these protein subunits are a central repetitive domain and non-repetitive terminal regions which contain most of the cysteine residues. N = N-terminal domain. C = C-terminal domain. S = cysteine residue

cross-linked matrix formed by intermolecular disulfide linkages among the HMW- and LMW-glutenin subunits (Schofield, 1986). Disruption of this network, such as by partial reduction of disulfide bonds, leads immediately to loss of dough functional properties (Miflin et al., 1983). A further confounding finding is that comparison of the genes of HMW-glutenin subunit genes correlated with good and poor quality reveals only minor differences in amino acid sequence. This may indicate that minor differences in sequence can have major functional consequences. Among the suggested sites for such dramatic quality effects are number and placement of cysteine residues and the number and order of peptide motifs in the repetitive domain. Thus, the study of these genes helps us understand the molecular basis of dough rheology, and suggests avenues of research to improve wheat quality and develop alternative utilizations for wheat.

Attempts to use the HMW-glutenin genes to bioengineer improved wheat quality and new utilizations will at first use native genes. In the simplest approach, new HMW-glutenin loci will be created *via* transformation. Studies have shown a correlation between wheat quality and the dosage of active HMW-glutenin genes (Lawrence et al., 1988). Cultivars expressing five gene tend to have better quality than those expressing four, which in turn have better quality than those expressing three, and so on. Thus there is a high expectation that quality can be improved with the addition of more HMW-glutenin genes. However, even if this is successful, the major impact will come once we understand enough of the molecular basis of the glutenin contribution to quality parameters to modify genes for enhanced, or even novel, glutenin characteristics.

Even with anticipated improvements in wheat transformation, it will not be feasible to test every possible modification to glutenin genes since the wheat flour functionality tests require enough mature seed to carry out physical studies and because generating transgenic wheat will still be too time-consuming. A desired alternative is that the protein products of novel gene constructs be testable without transformation. One approach to this problem is to express modified genes in microbial expression systems. The expressed glutenin can then be purified and studied in the laboratory. Such a strategy was shown to be feasible (Galili, 1989), and we are collaborating with several laboratories to use a bacterial expression system. HMW-glutenins purified from bacteria have been used to begin studies of the specificities of disulfide crosslinks, theunusual conformational characteristics of these proteins, and the *in vitro* formation of inter-molecular linkages (Shani et al., 1992). The purified proteins should also be useful in basic physical-chemical studies of these unique proteins.

Although the aforementioned studies are providing important information on the HMW-glutenins and have even greater potential in basic protein chemistry studies, they cannot address the effect of laboratory modifications on functionality of wheat; i.e., the viscoelastic characteristics of dough. For this we need a dough test system where microbially produced proteins can be introduced in "real-world" testing of dough parameters. The presentation at this meeting by Frank Bekes (CSIRO, Sydney) will describe just such a dough functionality testing system.

Finally, let us speculate on what a "complete" model of glutenin bioengineering might encompass (Fig. 3) At the center will be the laboratory modification of HMW-gluteningenes. Eventually some of these modified constructs will be transferred to wheat elite breeding lines to be used in developing new cultivars. Which modifications will be the most useful? The answer is that modified genes will first be tested in one or two model expression systems. One we have already discussed, i.e., bacterial expression systems. As we learn from the results of dough-testing and physical-chemical studies, we will understand what modifications are most likely to produce economically viable improvements in the glutenins. A second expression system could be "model" wheat lines that express either no HMW-glutenins or express only one or two well characterized subunits. New HMW-glutenin genes will be

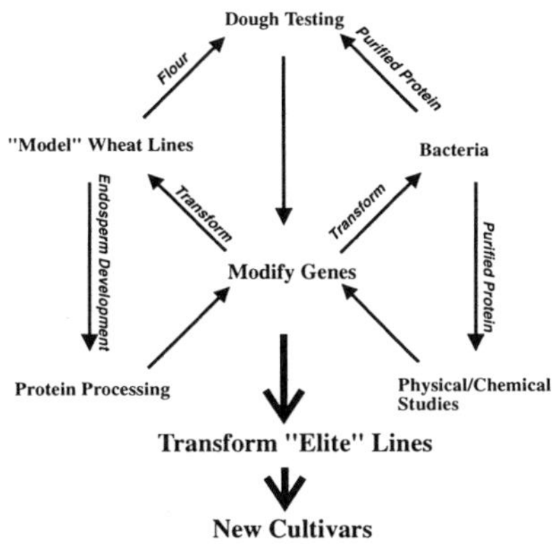

Figure 3. Representation of the research area interrelationships in a proposed "Complete Glutenin Bioengineering Program." A flowchart describes a view of how different research problems and approaches can be related to the overall goal of using the HMW-glutenin subunit genes to bioengineer wheat for improved and novel uses.

introduced into these lines which can then be used to study their effects on dough functionality. The new lines may form the basis of studies on endosperm protein processing and protein body formation. Again, the knowledge gained will feed back into further gene modifications. Through these cycles on gene modification and product analysis, we will learn how to alter the HMW-glutenin genes to not only enhance quality characteristics for known uses, but we will develop new utilizations not yet conceived.

CONCLUDING REMARKS

Progress in removing the technical barriers to wheat bioengineering raises new non-technical issues that may prove even more challenging to resolve. How will transformed cereal lines be integrated into traditional breeding programs? How will new products be commercially developed? How will the public react to products from bioengineered grains? As researchers, we find it encouraging that the availability of transformation technology and characterized gene systems now makes such questions pertinent.

REFERENCES

Anderson, O.D., Greene, F.C., Yip, R.E., Halford, N.G., Shewry, P.R. and Malpica-Romero, J.M. (1989) Nucleotide sequences of two high-molecular-weight glutenin subunit genes from the D-genome of a hexaploid bread wheat, *Triticum aestivum* L. cv Cheyenne. *Nucleic Acids Res.* 17:461–462.

Anderson, O.D., Halford, N. G., Forde, J., Yip, R. E., Shewry, P. R., and Greene, F. C. (1988) Structure and analysis of the high-molecular-weight glutenin genes from *Triticum aestivum* L. cv Cheyenne. Seventh Int. *Wheat Genetics Symposium Proc.* pp. 699–704. Bath Press, Cambridge, England.

Bower, R. and Birch, R.G. (1992) Transgenic sugarcane plants via microprojectile bombardment. *Plant Journal* 2:409–416.

Fromm, M.E., Morrish, F., Armstrong, C., Williams, R., Thomas, J., and Klein, T.M. (1990) Inheritance and expression of chimeric genes in the progeny of transgenic maize plants. *Bio/Technology* 8:833–839.

Gordon-Kamm, W.J., Spencer, T.M., Mangano, M.L., Adams, T.R., Daines, R.J., Start, W.G., O'Brien, J.V., Chambers, S.A., Adams, W.R., Jr, Willetts, N.G., Rice, T.B., Mackey, C.J., Krueger, R.W., Kausch, A.P., and Lemaux, P.G. (1990) Transformation of maize cells and regeneration of fertile transgenic plants. *Plant Cell* 2:603–618.

Hart, C.H., Fischer, B., Neuhaus, J.-M., and Meins, F.-Jr (1992) Regulated inactivation of homologous gene expression in transgenic Nicotiana sylvestris plants containing a defense-related tobacco chitinase gene. *Mol. Gen. Genet.* 235:179–188.

Lawrence, G.J., Macritchie, F., and Wrigley, C. W. (1988). Dough and baking quality of wheat lines deficient in glutenin subunits controlled by the *Glu-A1*, *Glu-B1* and *Glu-D1* loci. *J. Cereal Sci* 7:109–112.

McKnight, R.A., Shamay, A., Sankaran, L., Wall, R.J., and Hennighausen, L. (1992) Matrix-attachment regions can impart position-independent regulation of a tissue-specific gene in transgenic mice. *Proc. Natl. Acad. Sci. USA* 89:6943–6947.

Meyer, P., Heidmann, I., and Niedenhof, I. (1993) Differences in DNA-methylation are associated with a paramutation phenomenon in transgenic petunia. *The Plant Journal* 4:89–100.

Miflin, B.J., Field, J.M. and Shewry, P.R. (1983) "Cereal storage proteins and their effect on technological properties". Chapter 12 in *Plant Proteins,* Daussant, J., Mosse, J. and Vaughn, J., eds. pp. 255-319, Academic Press, New York.

Payne, P.I., Corfield, K.G., Holt, L.M., and Blackman, J.A. (1981) correlations between the inheritance of certain high-molecular-weight subunits of glutenin and bread making quality in progenies of six crosses of bread wheat. *J. Sci. Food Agric.* 32:51–60.

Payne, P.I. (1987) Genetics of wheat storage proteins and the effect of allelic variation on bread-making quality. Ann. Rev. *Plant Physiol.* 38:141.

Schofield, J. D. (1986) Flour Proteins: Structure and Functionality in Baked Products. in *Chemistry and Physics of Baking,* Blanshard, J. M. V., Fraizer, P. J. and Galliard, T.(Eds.), pp. 14-29. Royal Society of Chemistry, London.

Sears, R. G. and Deckard, E.L. (1982) Tissue culture variability in wheat: callus induction and plant regeneration. *Crop Sci.* 22:546–550.

Shani, N., Steffen-Campbell, J. D., Anderson, O. D., Greene, F. C., and Galili, G. (1992) Role of the Amino- and Carboxy-Terminal Regions in the Folding and Oligomerization of Wheat High Molecular Weight Glutenin Subunits. *Plant Physiol.* 98:433–441.

Shewry, P.R., Halford, N.G., and Tatham, A.S. (1992) High molecular weight subunits of wheat glutenin. *J. Cereal Sci.* 15:105–120.

Shimamoto, K., Terada, R., Izawa, T., and Fujimoto, H. (1989) Fertile transgenic rice plants regenerated from transformed protoplasts. *Nature* 338:274–276.

Somers, D.A., Rines, H.W., Gu, W., Kaeppler, H.F., and Bushnell, W.R. (1992) Fertile, transgenic oat plants. *Bio/Technology* 10:1589–1594.

Toriyama, K., Arimoto, Y., Uchimiya, H., and Hinata, K. (1988) Transgenic rice plants after direct gene transfer into protoplasts. *Bio/Technology* 6:1072–1074.

Urry, D. W. (1984). Protein elasticity based on conformations of sequential polypeptides: the biological elastic fiber. *J. Protein Chem.* 3:403–436.

Vasil, V., Castillo, A., Fromm, M., and Vasil, I. (1992) Herbicide resistant fertile transgenic wheat plants obtained by micro-projectile bombardment of regenerable embryogenic callus. *Bio/Technology* 10:667–674.

Weeks, J.T., Anderson, O.D., and Blechl, A. (1993) Rapid production of multiple independent lines of fertile transgenic wheat (*Triticum aestivum*). Plant Physiology 102:1077–1084.

Zhang, W. and Wu, R. (1988) Efficient regeneration of transgenic plants from rice protoplasts and correctly regulated expression of the foreign gene in the plants. *Theor. Appl. Genet.* 76:835–840.

THE CONTRIBUTIONS TO MIXING PROPERTIES OF 1D HMW GLUTENIN SUBUNITS EXPRESSED IN A BACTERIAL SYSTEM

F. Bekes,[1] O. Anderson,[2] P. W. Gras,[1] R. B. Gupta,[3] A. Tam,[2] C. W. Wrigley,[1] and R. Appels[3]

[1]CSIRO Division of Plant Industry
Grain Quality Research Laboratory
North Ryde, NSW 2113 Australia

[2]USDA Western Regional Research Center
800 Buchanan Street, Albany, California

[3]CSIRO Division of Plant Industry
Canberra, ACT 2600, Australia

INTRODUCTION

Strong correlative evidence indicates that specific subunits of the glutenin proteins are at least in part responsible for differences in dough properties (Payne, 1987). In particular, HMW subunits coded by the 1D chromosome have been found to be strongly associated with dough strength. The evidence for these findings is based on correlative studies. Therefore, the individual roles of these subunits in dough properties needs to be clarified by direct measurement. In the past, direct testing of the effects of specific proteins on dough structure has been possible only when large amounts of protein have been available. Conventional laboratory mixing procedures require 10–250 g flour and the isolation of purified proteins sufficient for testing on this scale is very difficult and expensive.

Two recent developments will greatly facilitate the study of these structure/function relationships: the development of micro-scale instruments testing dough properties and the expression of specific gluten proteins in heterologous systems. The development of the 2g Mixograph (Gras et al., 1990) and the associated automated interpretation of the resulting Mixograms provides results for mixing parameters which are effectively identical to those obtained from larger instruments (Rath et al., 1990). This has allowed mixing studies to be applied in situations where sample sizes are limited (Bekes and Gras, 1992). The expression of specific gluten proteins in heterologous systems provides the potential for the production of useful amounts of highly purified components of this otherwise very complex mixture of proteins. Furthermore, the use of this technique coupled with protein engineering will allow detailed study of the role of specific structural features of the polypeptides in dough properties (Madgwick et al., 1992). Investigation of the effects of the individual polypeptides on dough properties can be carried out by mixing studies in which the polypeptides are chemically incorporated into the proteins of the base flour. This technique has already been used to determine mixing properties of flour samples with glutenin composition modified by the addition/incorporation of the purified HMW glutenin subunit 1Bx20 (Bekes et al., 1994b).

This paper reports the investigation of the contribution of expressed 1D HMW glutenin subunits to functional properties by direct mixing experiments using a carefully controlled reduction/oxidation procedure in the 2 g-Mixograph.

MATERIALS AND METHODS

Medium-protein (MPF) and high-protein (HPF) commercial bakers' flours (10.3% and 13.0% protein, N X 5.7, respectively) were used as base flours in this study. In special cases, flour samples from a wheat line deficient for HMW glutenin subunits 5+10 (1,17+18,-) or a sibling with normal HMW glutenin subunit composition (1,17+18,5+10) was used as base flour. These lines were present in Gabo/Olympic mixed backgrounds (Lawrence et al., 1988) and were grown at Narrabri in 1990.

Colonies of the *E. coli* strain BL21 (DE3) pLys, harbouring expression plasmid based on the pET-3a vector (Rosenberg et al., 1987). Coding sequences of HMW glutenin subunits 1Dx2, 1Dx5, 1Dy10 and 1Dy12 have been cloned into this vector. Plasmids were grown in ZY medium (Shani et al., 1992) and HMW glutenin subunit synthesis was induced for polypeptide accumulation and isolation. Following a lysozyme treatment and centrifugation, proteins were extracted from the bacterial pellets with 70% (v/v) aqueous ethanol at 20°C and the supernatant mixed with 2 volumes of 1.5 M aqueous NaCl and left overnight at 4°C to precipitate. After centrifugation, the pellet was dispersed in 0.1M acetic acid, recentrifuged and the supernatant freeze dried.

Mixing tests were conducted with a prototype 2 g-Mixograph using a modification of the standard method for 35 g flour scaled to the two gram size (AACC, 1983). Mixing parameters were determined using a modification of a previously reported computer program. This modification automatically excised the portions of the recording during which mixing was halted. A reversible reduction/oxidation procedure was used for incorporating glutenin subunits into glutenin. Flour (2.00 g HPF or MPF) with or without purified proteins (2–7 mg) were mixed with 1.00 mL water and 0.10 mL water containing 50 μg/mL DTT for 30 seconds and allowed to react for four minutes. The reduced doughs were then treated with 0.10 mL oxidant solution containing 200 μg/mL of potassium iodate. Mixing was continued for 30 seconds and the dough then allowed to react for 5 minutes and then mixed for a further ten minutes, as in a conventional Mixograph determination ('incorporation'). The same protocol was followed in separate experiments but with 0.10 mL water additions instead of the reducing and oxidizing agents ('simple addition'). Subunits were added/incorporated singly or in combinations of specific pairs of subunits. All mixing tests were performed in duplicate. Molecular weight distributions of dough proteins were determined from the total protein extracts (Singh et al., 1990), using SE-HPLC (Batey et al., 1991). Protein fractions were isolated from the SE-HPLC by collecting fractions corresponding to selected MW size ranges. The subunit compositions of these fractions were characterised by SDS-PAGE (Gupta and MacRitchie, 1991).

RESULTS AND DISCUSSION

The generally accepted molecular structure of glutenin proteins implies that the effects of any particular subunit on dough properties are not exerted by the monomeric subunit, but rather, as a contribution to the structure of the glutenin polymer. Therefore, meaningful estimates of the effects of added glutenin subunits on dough properties could be made only if they can be chemically incorporated into the glutenin polymer. A technique has recently been developed for incorporating glutenin subunits into the dough structure by partial reduction of a mixture of a specific subunit and a base flour, followed by reoxidation (Bekes et al., 1994a).

Glutenin proteins of mixed doughs were partially resolved into two components, P_I, consisting of large polymeric glutenin and P_{II}, smaller polymeric glutenin and monomeric HMW subunits (Figure 1). After reduction/oxidation treatment, dough samples containing

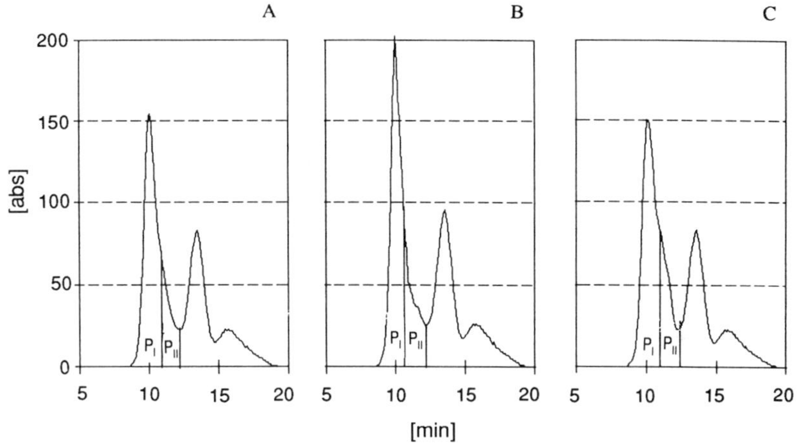

Figure 1. SE-HPLC separations of total proteins extracts from dough samples. A: MPF flour; B: MPF flour with 1D 5 HMW subunit incorporated; C: MPF flour with addition of 1D 5 HMW subunit.

expressed HMW glutenin subunits showed a significant increase in the amount of P_I with a slight increase in the amount of P_{II}. In contrast, no changes in the amount of P_I and a significant increase in the amount of P_{II} were found in dough samples containing the same amount of subunits prepared by simple addition. Fractions corresponding to the SE-HPLC peaks P_I and P_{II} were collected, freeze dried and characterised by SDS-PAGE. With incorporation, subunits were found predominantly in P_I, but with simple addition, they were found predominantly in P_{II}. These observations indicate that the expressed subunits have been incorporated into the glutenin structure by the reduction/oxidation procedure. Densitometric analysis of SDS-PAGE separations of reduced proteins from P_I and P_{II} fractions showed that less than 10% of the subunits used in simple addition experiments were spontaneously incorporated into the glutenin structure, indicating only limited sulfhydryl-disulfide interchange during mixing.

The incorporation of 1D subunits 2, 5, 10 or 12 led to greater mixing requirements (increased mixing time and peak resistance) and increased tolerance to overmixing (decrease in resistance breakdown). These effects were found to be consistent in both base flours used, but changes were slightly higher in the case of MPF, presumably because of the lower protein content. The incorporation of any of these HMW subunits increased the strength of the dough, as predicted from earlier studies (Tatham et al., 1991). In contrast, simple addition of subunit polypeptides resulted in the opposite effects—shorter mixing requirement with a more rapid breakdown after maximum resistance. The nature of the latter changes is similar to those observed when the gliadin/glutenin ratio of flours has been modified by gliadin addition (Gras et al., 1992).

The diametrically opposite effects of incorporated and added subunits on mixing properties indicate that the size of the subunit present in its monomeric form does not strengthen dough characteristics. The real contribution of the polypeptide to the structure/function relationship can be estimated only if it is incorporated into the glutenin polymer, using a reduction/oxidation system.

The Effect of Incorporating 5+10 or 2+12 Pairs

A flour lacking the 1D-coded subunits (5+10) or (2+12) was supplemented by the incorporation of 5mg of a 1:1 molar ratio of the expressed subunits (5+10) or (2+12). Comparison of the mixing results of these trials with those of the sibling line containing the subunits (5+10) in the same background showed (Figure 2).

1. Mixing curves from experiments where (5+10) subunits were incorporated and from native flour containing (5+10) were effectively identical.
2. Doughs obtained by incorporation of (2+12) were not as strong as those containing an equivalent amount of (5+10).

In a separate experiment, varying molecular ratios of the 1D subunit pairs (5+10) or (2+12) were incorporated into both MPF and HPF flours. It was found that the response of the mixing properties was always higher when both components were present. This synergistic effect was at a maximum when equimolar ratios of either 5 and 10 or 2 and 12 were used (Figure 3).

The Effect of Size of 1D Subunits

For each of the alleles coded by the Glu-1D gene, only one x-type subunit (2 or 5) and only one y-type subunit (10 or 12) are expressed in wheat grain. In nature, the subunit pairs (5+10) and (2+12) are always expressed together. The x- and y-type proteins differ in that x-type subunits have higher MW and the y-type subunits have more cysteine residues which are available for crosslinking.

Having each of these subunits available allowed the testing of un-natural x-y pairs of subunits such as (2+10) and (5+12) and even x-x (2+5) and y-y(10+12) pairs. When MPF flour was enriched with 5 mg of single 1D subunits or a 1:1 molar ratio of any of the possible pairs of 2, 5, 10 and 12 subunits, it was found that the response of x-type subunits or pairs of x-type subunits was significantly larger than from y-type subunits or pairs of y-type subunits. The highest responses were produced by pairs consisting of one x- and one y-type subunit, regardless of whether it was a natural combination or not (Figure 4).

The data from the incorporation of single subunits into the glutenin show a significant positive correlation between subunit MW and mixing time (Figure 5). .However, incorporation of x-y pairs of subunits shows responses much higher than expected from their calculated average molecular weights. This synergistic effect indicates a special interaction between x-type and y-type 1D subunits, which must play an important role in determining the structure/function relationships of these proteins.

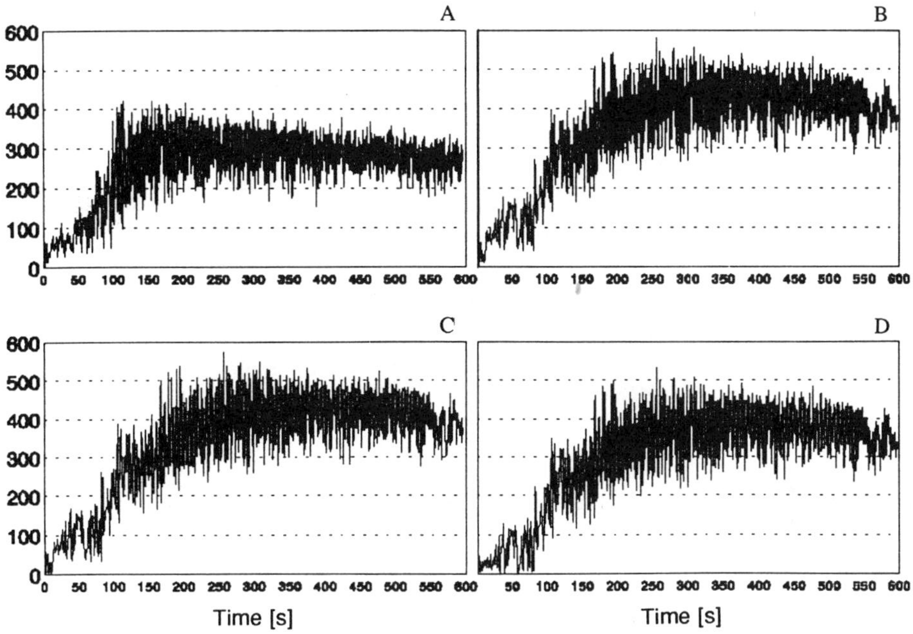

Figure 2. The effects on mixing properties caused by incorporating mixtures of 1–1 mg of 1Dx5+1Dy10 (B) or 1Dx2+1Dy12 (C) into the glutenin of a (1,17+18,-) line (A). Results for mixing the sibling line with normal HMW glutenin subunit composition (1,17+18,5+10) are shown as (D).

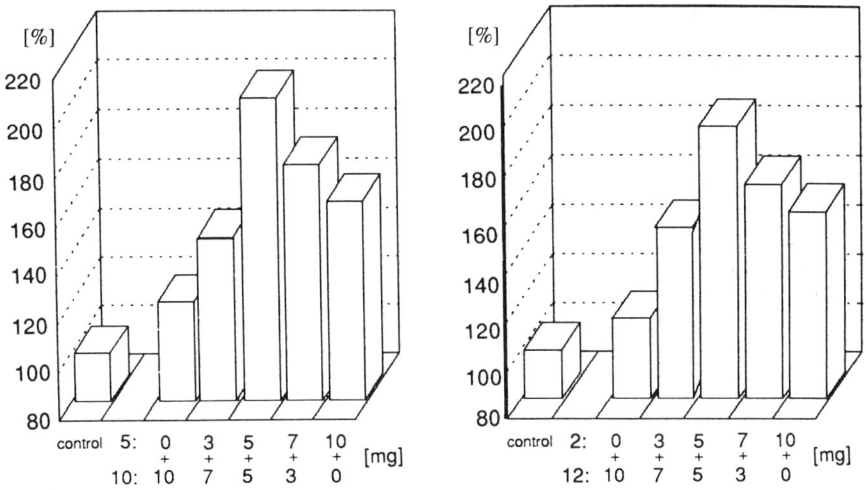

Figure 3. The effects on mixing time of MPF flour caused by incorporating 5+10 or 2+12 subunits in different ratio. (Results are expressed as the percent of control.)

All of these results relate to polypeptides expressed in *E. coli*. We have not yet established that these polypeptides are identical to the native polypeptides in amino-acid sequence

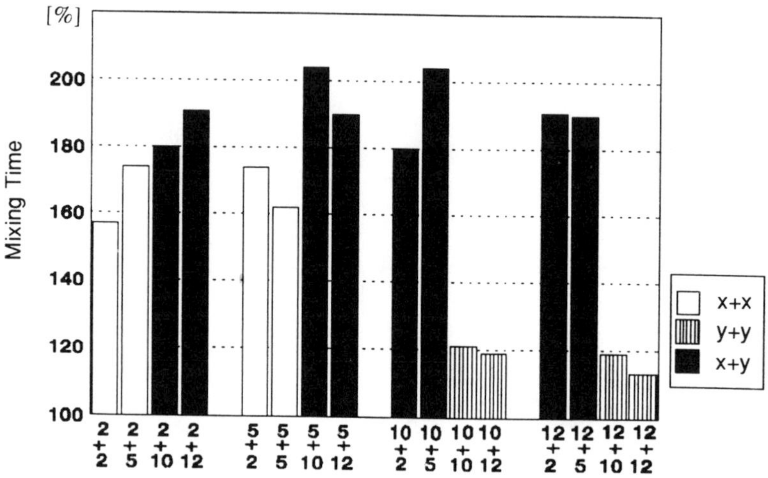

Figure 4. The effects on mixing time of MPF flour caused by incorporating 5 mg of 1:1 mixtures of 1D HMW glutenin subunits.

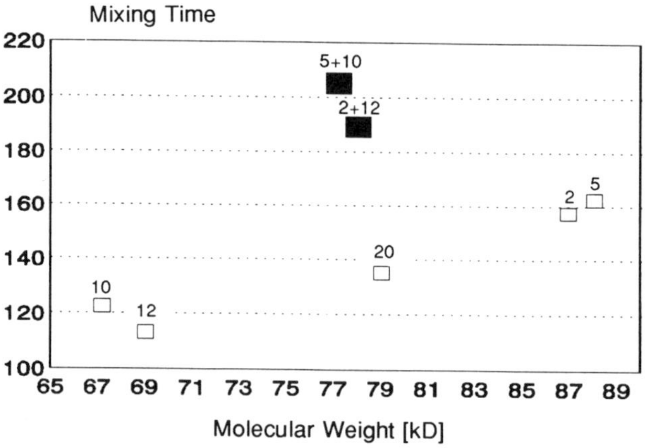

Figure 5. The relationship between the molecular weight of incorporated HMW glutenin subunits and their effect on mixing time (in seconds), including results for 1By20 subunit purified from wheat (Bekes et al., 1993b).

and conformation. Nevertheless, where comparisons with native proteins were made, the expressed polypeptides had very similar functionality to native subunits.

CONCLUSION

The individual contributions of expressed HMW glutenin subunits to functional properties have been investigated for the first time. The results of direct mixing experiments using purified subunits and a carefully controlled reduction/oxidation procedure on the two-gram Mixograph showed significant changes in functional properties when expressed HMW glutenin subunits were incorporated into the structure of glutenin. Simple addition of these subunits to doughs did not improve dough strength.

A synergistic effect on mixing properties was observed when x- and y-types subunits were incorporated together. The maximum effect was observed when close to the 1:1 molar ratio of the subunits were incorporated.

When (5+10) subunits were incorporated into a flour lacking the 1D-coded subunits the mixing properties were almost identical to those of a flour from a sibling line where the natural composition of glutenin included (5+10) subunits. As expected, doughs obtained by incorporation of (2+12) subunits were not as strong as those containing an equivalent amount of (5+10) subunits.

Direct mixing results in this study highlight the importance of the size of the HMW glutenin subunits on mixing properties. It has also been observed that certain combinations of these subunits caused larger effects than could be assumed from the size effect only. This implies that specific interactions between x- and y-type subunits play an important role in determining protein structure/function relationships.

ACKNOWLEDGEMENT

We wish to acknowledge Dennis Murray for expert technical assistance.

REFERENCES

American Association of Cereal Chemists (1983) 'Approved Methods,' AACC, St Paul, Minnesota Method No. 54–40.

Batey, I.L., Gupta, R.B., and MacRitchie, F. (1991) *Cereal Chem.* 68, 207–209.

Bekes, F., and Gras, P.W. (1992) *Cereal Chem.* 69, 229–230.

Bekes, F., Gras, P.W., and Gupta, R.B. (1994a) Cereal Chem. 71, 44–50.

Bekes,F., Gras P.W., Gupta R.B., Hickman, D.R., and Tatham, A.S. (1994b) J. Cereal Sci. 19, 3–7.

Gras, P.W., Bekes, F., Gupta, R.B. and MacRitchie, F. (1992) Proc. 42nd RACI *Cereal Chem. Conf.* (Humphrey-Taylor, V.J. ed), RACI, Parkville, Australia, pp. 171–180.

Gras, P.W., Hibberd, G.E., and Walker, C.E. (1990) Cereal Foods World 35, 568–571.

Gupta, R. B. and MacRitchie, F.(1991) J. Cereal Sci. 14, 205.

Lawrence, G., MacRitchie, F., and Wrigley, C.W. (1988) J. Cereal Sci.8 109–112.

MacRitchie, F. (1987) J. Cereal Sci. 6, 259–266.

Madgwick, P., Pratt, K., and Shewry, P.R. (1992) Expression of wheat gluten proteins in hetrologous systems. In: *Plant Protein Engineering.* Eds. Shewry, P.R. and Gutteridge, S., pp. 188–200., Cambridge University Press, Cambridge.

Payne, P.I. (1987) *Annu. Rev. Plant Physiol.* 38, 141–157.

Rath, C.R., Gras, P.W., Wrigley, C.W., and Walker, C.E. (1990) *Cereal Foods World* 35, 572–574.

Rosenberg, A.H., Lade, B.N., Chui, D., Lin, S., Dunn, J.J., and Studier, F.W. (1987) *Gene* 56 125–135.

Shani, N., Steffen-Campbell, J.D., Anderson, O.D., Green, F.C., and Galili, G. (1992) *Plant Physiol.* 98, 433–441.

Singh, N.K., Donovan, G.R., Batey, I.L., and MacRitchie, F. (1990) *Cereal Chem.* 67, 150–161.

Tatham, A.S., Field J.M., Keen, J.N., Jackson P.J., and Shewry, P.R. (1991) *J. Cereal Sci.* 14, 111–116.

STUDIES OF HIGH MOLECULAR WEIGHT GLUTENIN SUBUNITS AND THEIR ENCODING GENES

D. Lafiandra,[1] R. D'Ovidio,[1] and B. Margiotta[2]

[1]Department of Agrobiology and Agrochemistry
University of Tuscia
Viterbo, Italy

[2]Germplasm Institute
C.N.R.
Bari, Italy

INTRODUCTION

Our knowledge of the biochemical and genetical aspects of high molecular weight (HMW) glutenin subunits has been possible through the extensive use of electrophoretic techniques and of SDS-PAGE in particular; in fact by using this technique the existence of allelic variation at each of the three *Glu-1* loci encoding HMW glutenin subunits in bread wheat and the influence they have in determining qualitative properties have been firmly established (Payne, 1987). Use of different and more sophisticated biochemical techniques, such as reversed-phase high performance liquid chromatography (RP-HPLC) or molecular studies conducted with new tools, such as the polymerase chain reaction (PCR) and nucleotide sequence analysis, are constantly supplying more information on these proteins and their encoding genes.

SDS-PAGE OF HMW GLUTENIN SUBUNITS

Although SDS-PAGE has been very useful for the study of HMW glutenin subunits some limits exist in the use of this technique as have been highlighted in many studies. Gene cloning and sequencing of HMW glutenin subunit genes have produced more detailed information on corresponding subunits and made it possible to identify discrepancies between the migration of certain subunits in SDS-PAGE and their molecular weight. For instance the migration of allelic pairs 1Dx2/1Dx5 and 1Dy10/1Dy12 have been shown to be anomalous (Greene et al. 1988). In fact subunit 1Dx5 has higher mobility than the smaller allelic subunit

Improvement of Cereal Quality by Genetic Engineering, Edited by
Robert J. Henry and John A. Ronalds, Plenum Press, New York, 1994

1Dx2; similarly subunit 1Dy10 has lower mobility than the larger subunit 1Dy12. It has been reported that, in the presence of a strong denaturant such as urea, HMW glutenin subunits display relative electrophoretic mobilities in accordance with that expected from their size differences and indeed this has been observed to be the case for the pairs of subunits 1Dx2/1Dx5 and 1Dy10/1Dy12 (Goldsbrough et al. 1989),but anomalies can still be found for other subunits. Lafiandra et al. (1993) reported that subunit 2 had a lower electrophoretic mobility than subunit 1, when separated on SDS-PAGE containing 4M urea, though subunit 2 is smaller than subunit 1 as DNA sequencing data have indicated (Sugyama et al. 1985; Halford et al. 1992).

In addition to exhibiting discrepancies between molecular weight and migration, the molecular weights of HMW glutenin subunits, as determined by their migration of SDS-PAGE, appear to be overestimated (Bunce et al. 1985). Further complications in interpreting SDS-PAGE data can arise from allelic components possessing identical apparent molecular weights. Examples of these subunits are reported in Fig. 1; subunits indicated as 4 and 4_1, present, respectively, in the bread wheat cultivars Newton and Kador (lanes 5 and 6) and 5^* and 5, from cultivars fiorello and Axona (lanes 7 and 13), cannot be distinguished between on SDS-PAGE at a 10% gel concentration. These complications might change the interpretation of qualitative data as will appear in other parts of this presentation.

RP-HPLC SEPARATION OF HMW GLUTENIN SUBUNITS

RP-HPLC has offered a new tool for the study and characterization of HMW glutenin subunits. One of the major features of RP-HPLC is its capability to quantitate different proteins more accurately compared to densitometric analyses of stained gels. Marchylo et al. (1989) reported that the relative proportion of subunit 7 present in two Canadian wheat cultivars were significantly different; these subunits also appeared to have slightly different mobilities on gradient SDS-PAGE, and have been indicated by them as 7 and 7^* (Marchylo et al. 1992). This is also the case for subunit 8, such as the one present in Kador and Newton in Fig. 1. Apparently there is no difference on SDS-PAGE between subunit 8 of these two cultivars however differences are detected when the HMW glutenin subunits are analysed by RP-HPLC. In fact, the subunits 8 present in Kador and Newton, differ in surface hydropho-

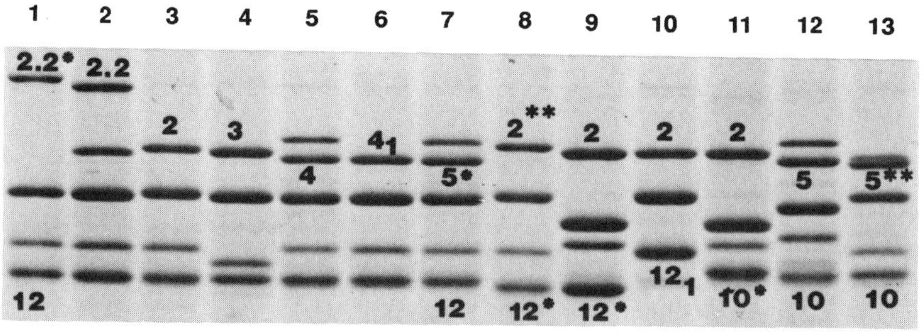

Figure 1. SDS-PAGE separation of HMW glutenin subunits present in lines and cultivars of bread wheats. Numbers of reported indicate allelic variants at the *Glu-D1* locus.

bicity, with that of the former being more hydrophobic and produced in smaller quantities than the latter.

Allelic variation of HMW glutenin subunits has been also studied. Recently Margiotta et al. (1993) have analysed several durum and bread wheat cultivars displaying a large number of variants at each of the three *Glu-1* loci. Allelic variants present at the *Glu-A1* and *Glu-D1* loci were characterised by small variations in surface hydrophobicity, whereas those present at the *Glu-B1* locus had a boarder range of hydrophobicities. RP-HPLC separation of HMW glutenin subunits has been carried out mostly on reduced and alkylated subunits, using 4-vinylpyridine as alkylating agent. Nevertheless, comparison of the variation observed in surface hydrophobicity when the same subunit is reduced or reduced and alkylated revealed additional information (Margiotta et al. 1993).

In fact, the introduction of pyridylethyl (PE) group to the various cysteine residues modifies surface hydrophobicity and, since separation is carried out under acidic conditions, PE-glutenin subunits are positively charged. This results in a reduction of elution time when a subunit is reduced and alkylated compared to the same subunit only reduced. Reduction in elution time following alkylation differentially affects proteins encoded at different loci and on x - or y-type subunits.

Table 1 reports retention times of alkylated and non-alkylated subunits; the reduction in elution time of alkylated subunits, measured as % of retention time of reduced subunit,

Table 1. Retention Times (min.) of HMW Glutenin Subunits Reduced or Reduced and Pyridylethylated

Luci	Alleles	Reduced	Reduced Alkylated	Difference	%	Cysteine number*
Glu-Dly	12	42.0	16.9	25.1	59.8	7[a]
	10	42.5	17.1	25.4	59.8	7[b]
Glu-Bly	9	71.5	26.2	45.3	63.4	7[c]
Glu-DIx	2	73.3	30.1	43.2	58.9	4[d]
	5	75.3	29.4	45.9	61.0	5[b]
Glu-BIx	7	67.9	35.9	32.0	47.1	4[e]
	17	68.9	36.5	32.4	47.0	4[f]
Glu-Alx	2*	60.0	45.8	14.2	23.7	4[e]
	1	60.8	46.4	14.4	23.7	4[g]

*Number of cysteine residues were deduced from corresponding sequenced genes: a. Thomson et al. (1985): b. Anderson et al (1989): c. Halford et al. (1987): d. Sugyama et al. (1985): e. Anderson and Greene (1989): f. Reddy and Appels. (1993): g. Halford et al. (1992)

was larger for y-type subunits which possess 7 cysteine residues, compared to x-type residues which possess 4 cysteine residues; subunit 5 that possesses 5 cysteine residues has a % of reduction lager compared to other Dx-type subunits. Margiotta et al. (1993) have reported that some subunits, such as 20 or 6 encoded at the *Glu-B1* locus, were subject to minor reductions in elution time when alkylated; they suggested that this could be caused by a lower number of cysteine residues present in these subunits.

RP-HPLC separation has also enable the detection of greater heterogeneity than that found with electrophoretic techniques. Subunits with apparently identical molecular weight can be shown to be different when analysed by RP-HPLC, as shown with subunits 8 of Kador and Newton. A further example is represented by subunits 4 present in the same cultivars; when separated by RP-HPLC, these two subunits appear to be the same on SDS-PAGE, but show different values of surface hydrophobicity as deduced by their different elution times (Fig. 2). A particular case is represented by the wheat cultivar Fiorello; this cultivar was reported to possess the pair of subunits 5+12 at the *Glu-D1* locus as a consequence of intralocus recombination (Pogna et al. 1987); RP-HPLC analyses of HMW glutenin subunits present in the bread wheat cultivar Fiorello have indicated that subunit 5 present in this cultivar has a different elution time compared to the subunit 5 usually found associated with subunit 10 and was indicated as 5^* (Margiotta et al. 1993). The RP-HPLC behaviour of subunit 5^* suggests that this subunit strongly resembles allelic subunit 2.

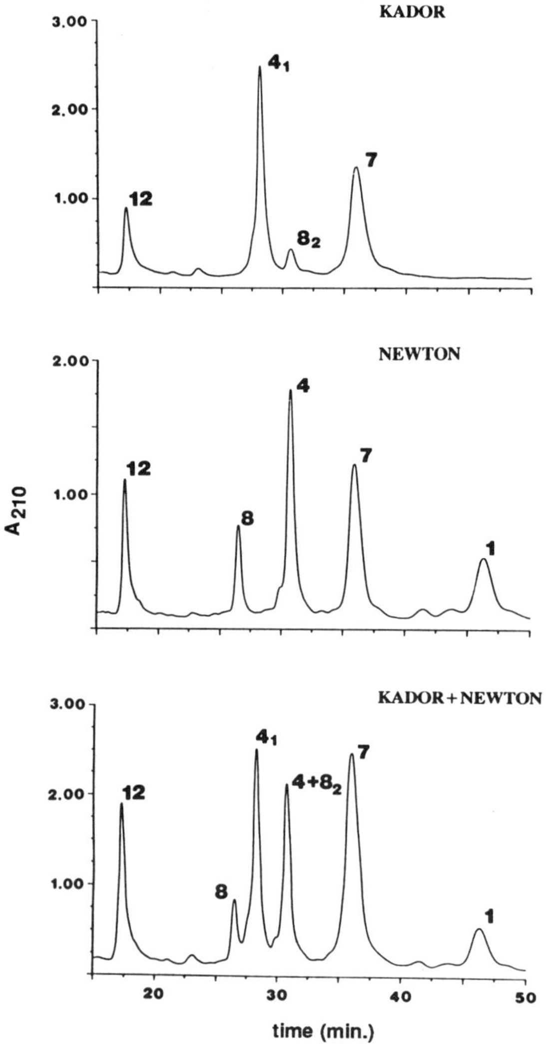

Figure 2. RP-HPLC separation of alkylated HMW glutenin subunits present in the bread wheat cultivars Kador, Newton and their mixture.

PCR: A NEW TOOL FOR STUDYING HMW GLUTENIN SUBUNITS

In the few years since its introduction, the polymerase chain reaction has become a widespread research technique, making it possible to further characterize different allelic forms and study their genetic polymorphism. The PCR approach, which allows the specific application of a target DNA segment, using a pair of flanking oligonucleotides as primers, has also been suggested as a valid alternative to the classical approach to isolate and characterize wheat storage protein genes (D'Ovidio et al. 1991; 1993a); what is more PCR allows identification of cultivars with different qualitative characteristics. For instance durum wheat cultivars with good and poor technological properties were distinguished by D'Ovidio (1993) by selective application of gamma gliadin genes or low-molecular weight glutenin subunit genes. But a more striking example is represented by the possibility of distinguishing bread wheat genotypes with genes encoding subunit 5 from those with subunit 2. In fact, the availability of nucleotide sequences of the HMW glutenin genes corresponding to subunits 2 and 5 has allowed specific oligonucleotides to be prepared for selective application of the latter (D'Ovidio and Anderson, 1993). Hence the problems encountered in SDS-PAGE separation have been overcome. In fact wheat cultivar Fiorello did not show any application band when its DNA was subjected to PCR analysis, further confirming the difference between its subunit and subunit 5, and its similarity to subunit 2. D'Ovidio et al. (1993b) have also used PCR to selectively amplify and measure the size of DNA fragments corresponding to allelic subunits present at the *Gly-D1* locus. The molecular weights of corresponding subunits were deduced and results indicated that problems such as overestimation or anomalous migration had been eliminated.

DNA sequencing studies have shown that x-type subunits contain three cysteine residuesin the non-repetitive N-terminal region and an additional cysteine in the C-terminal domain; the internal repeat regions do not contain additional cysteine residues, except for subunit 5, which has an extra cysteine residue at the beginning of the repetitive domain adjacent to the N-terminal region. PCR was used by Lafiandra et al. (1993) to selectively amplify and clone the N-terminal region of subunit 5 of Fiorello, to further characterize this gene. Results of these analyses indicated the absence of the characteristic cysteine residue, typical of subunit 5, in the DNA amplified fragment of Fiorello.

CONCLUSIONS

Electrophoretic separation has greatly contributed to our knowledge of HMW glutenin subunits; a combination of approaches such as those described will make an even greater contribution to the clarification of their role in the breadmaking properties of flours and the avoidance of misleading results such as those presented. The finding that subunit 5 of Fiorello is more related to subunit 2 and very likely possesses the same number of cysteine residues as subunit 2 reopens the question on the relative importance of the x-and y-type subunits at the *Glu-D1* locus. In fact, following the detection of the recombinant pair 5+12 in Fiorello Pogna et al. (1987) could not find any differences in the SDS-sedimentation values between lines carrying the pairs 5+12 and 2+12 obtained from crosses between Fiorello and cultivars possessing either the allelic pair 5+10 or 2+12. In contrast, significant differences were found when lines carrying 5+12 and 5+10 were compared leading to the conclusion that they y-type subunits 10 or 12 were the main components responsible for differences in the technological characteristics of flour. From results obtained by Margiotta et al. (1993) and Lafiandra et al. (1993), it appears clear that Pogna et al. (1987) have, in their crosses, actually compared 5^*+12 with 2+12 rather than 5+12 with 2+12. The lack of differences in the latter comparison may then be ascribed to the close similarity between 5^* and 2, whereas differences found in the first cross may be associated with differences existing between 5^* and 5. These results favour the hypothesis that the additional cysteine residue present in subunit 5, compared to subunit 2, is more likely to be responsible for superior qualitative performance observed in the pairs 5+10 vs 2+12 (Greene at al. 1988) through the formation of a greater proportion of larger sized polymeric glutenins (Gupta and MacRitchie, 1994).

RP-HPLC, by allowing the identification of heterogeneity not detectable by SDS-PAGE, along with more precise quantitation of HMW glutenin subunits, is clarifying apparently contrasting results previously reported using electrophoretic procedures. For instance Marchylo et al. (1992) have reported that quality scores assigned to HMW glutenin subunit pairings containing subunit 7 vary from 1 to 3, but separation of this subunit into two types, 7 and 7^* differing in their relative proportion, might be responsible for different scores; in fact the proportion of subunit 7 relative to the total amount of HMW glutenin subunits was significantly higher than for subunit 7^* and this seemed to be associated with greater dough strength (Marchylo et al. 1992). Similarly the heterogeneity observed by RP-HPLC for subunit 8 might influence differently the breadmaking performance of flours.

REFERENCES

Anderson, O.D., Greene, F.C., Ryan, E.Y. Halford, N.G., Shewry, P.R. and Malpica-Romero, J-M. *Nucleic Acids Res.* 17 (1989) 461–462.

Anderson, O.D., and Green, F.C. *Theor. Appl. Genet.* 77 (1989) 689–700.

Bunce, N., White, R.P. and Shewry, P.R. *J. Cereal Sci.* 3(1985) 131–142.

D'Ovidio, R. Plant Mol. Biol. 22 (1993) 1173–1176.

D'Ovidio, R., Anderson, O.D., and Porceddu, E. In 'Proceedings of the 5th International Workshop on Gluten Proteins.' (1993a) in press.

D'Ovidio, R., Tanzarella, O.A. and Porceddu, *E. Plant Sci.* 75 (1991) 229–236.

D'Ovidio, R. and Anderson, O.D. Theor. Appl. Genet. (1993b) in press.

D'Ovidio, R., Anderson, O.D. and Poceddu, E. In: 'Proceedings of the 5th International Workshop on Gluten Proteins' (1993a) in press.

Greene, F.C., Anderson, O.D., Yip, R.E. Halford, N.G., Malpica-Romero, J-M. and Shewry, P.R. Proc. 7th Int. Wheat Genet. Symp. IPSR, Cambridge, (1988) 735–740.

Goldsbrough, A.P., Bulleid, N.J., Freedman, R.B. and Flavell, R.B. *Biochem. J.* 263 (1989) 837–842.

Gupta, R.B. and MacRitchie, F.J. Cereal Sci. 19 (1994) 19–29.

Halford, N.G., Forde, J., Anderson, O.D., Greene, F.C. and Shewry, P.R. *Theor. Appl. Genet.* 75 (1987) 117–126.

Halford, N.G., Field, J.M., Blair, H., Urwin, P., Moore, K., Robert, L., Thompson, R.D., Flavell, R.N., Tatham, A.S. and Shewry, P.R. *Thor. Appl. Genet.* 83 (1992) 373–378.

Lafiandra, D., D'Ovidio, R., Porceddu, E., Margiotta, B. and Colaprico, G. J. Cereal Sci, 18 (1993) 197–205.

Marchylo, B.A., Kruger, J.E. and Hatcher, D.W.J. *Cereal Sci.* 9 (1989) 113–130.

Marchylo, B.A., Lukow, O.M. and Kruger, J.E. J. *Cereal Sci.* 15 (1992) 29–37.

Margiotta, B., Colaprico, G., D'Ovidio, R. and Lafiandra, D. *J. Cereal Sci.* 17 (1993) 221–236.

Margiotta, B., Colaprico, G., Turchetta, T. and Lafiandra, D. In: 'Proceedings of the 5th International Workshop on Gluten Protein'' (1993) in press.

Payne, P.I. *Ann. Rev. Plant Physiol* 38 (1987) 141–153.

Pogna, N.E., Mellini, F., and Dal Belin Peruffo, A. In *'Hard wheat: Agronomic, Technological, Biochemical and Genetical Aspects'* (B. Borghi, ed.), CEC, Luxembourg (1987) 53–69.

Reddy, P. and Appels, R. *Theor. Appl. Genet.* 85 (1993) 616–624.
Sugyama, T., Rafalsky, A., Peterson, D. and Soll, D. *Nucl. Acids Res.* 13 (1985) 8729–8737.
Thompson, R.D., Bartels, D. and Harberd, N.P. *Nucleic Acids Res.* 13 (1985) 6833–6846.

SECTION III

GENETIC ENGINEERING OF CEREAL STARCH QUALITY

PROSPECTS FOR THE PRODUCTION OF CEREALS WITH IMPROVED STARCH PROPERTIES

Jack Preiss,[1] David Stark,[2] Gerard F. Barry,[2] Han Ping Guan,[1] Yael Libal-Weksler,[1] Mirta N. Sivak,[1] Thomas W. Okita,[3] and Ganesh M. Kishore[2]

[1]Department of Biochemistry
Michigan State University
East Lansing, Michigan 48824

[2]Plant Sciences Technology
The Agricultural Group
The Monsanto Company
St. Louis, Missouri 63198

[3]Institute of Biological Chemistry
Washington State University
Pullman, Washington 99164

SUMMARY

The dominant pathway for the synthesis of starch involves three enzymes; ADPglucose pyrophosphorylase (ADPGlc PPase; EC 2.7.7. 27), which catalyzes the synthesis of ADPglucose; starch synthase (EC 2.4.1.21), which transfers the glucosyl portion of ADPglucose to a maltodextrin primer for synthesis and elongation of the α-1,4 glucosyl chain and the branching enzyme (EC; 2.4.1.18) which transfers a portion of the elongated α-1,4 glucosyl chain to form the α-1,6 branch points present in amylopectin (and to small extent, in amylose). The physical properties of the starch in a plant are dependent on the properties and catalytic activities of the three enzymes mentioned above and alteration of the enzyme amounts and their properties will in turn, affect the properties of the starch synthesized. Recent results demonstrate that transformation of certain plants with a bacterial ADPGlc PPase gene can dramatically increase their starch content. It is quite possible that alteration of the proportion or amounts of branching enzyme and/or starch synthase would alter the structure and physical properties of the starch synthesized. To this end, the purified branching isoenzymes and soluble starch synthase isozymes are characterized to determine their specific roles in the synthesis of amylopectin. Preliminary experiments suggest that maize

Improvement of Cereal Quality by Genetic Engineering, Edited by
Robert J. Henry and John A. Ronalds, Plenum Press, New York, 1994

BE I isoenzyme is mainly involved in synthesis of the B chains of amylopectin while BE IIa and IIb are involved in the synthesis of the A chains.

INTRODUCTION

The ADPglucose Pathway

According to present understanding, the key enzymes of starch biosynthesis are:

1. ADPglucose pyrophosphorylase (ADPGlc PPase), which catalyzes synthesis of the glucosyl donor ADPglucose;

$$\alpha\text{-glucosyl-1-P} + \text{ATP} \Longleftrightarrow \text{ADPglucose} + \text{PP}_i$$

2. starch synthase, which transfers the glucosyl unit of ADPglucose to the non-reducing end of an αα 1,4-glucan primer;

$$\text{ADPglucose} + (\text{glucosyl})_n \longrightarrow \text{ADP} + (\text{glucosyl})_{n+1}$$

and

3. the branching enzyme, which forms the α-1,6-linkage found in amylopectin and to a small extent in amylose.

linear glucosyl chain of α-glucan ------> Branched chain of α-glucan with α-1->6 linkage branch point.

Many biochemical and genetic studies support the view that the ADPglucose pathway involving the three reactions mentioned above is the main or only one resulting in starch synthesis. For example, studies of isolated mutants of maize endosperm (Tsai and Nelson, 1966; Dickinson and Preiss, 1969) deficient in ADPGlc PPase activity are also deficient in starch content. The *rb* locus controls the level of ADPGlc PPase levels in developing pea embryos (Smith et al., 1989). One pea line having recessive *rb* genes contained 3–5% of the ADPGlc PPase activity and 38–72% of the starch found in the pea line having the dominant genes. Moreover, an *Arabidopsis thaliana* mutant that contains less than 2% of the starch levels seen in the normal line has less than 2% of the normal ADPGlc PPase levels (Lin et al, 1988). Confirming the information obtained with ADPGlc PPase deficient mutants, Müller-Röber and colleagues (1992), expressed in potato tuber a chimeric gene encoding antisense RNA for the ADPGlc PPase, causing a reduction in enzymatic activity to 2 to 5% of the wild type levels and a commensurate reduction in starch content. Other data demonstrating a direct relationship between increase in activity of the above starch biosynthetic enzymes and increased starch accumulation in various plants have been previously reviewed (Okita, 1992; Preiss, 1988; 1991; Preiss and Levi, 1980).

The kinetic properties of the enzymes in the ADPglucose pathway (K_m and V_{max} values) together with the concentrations of substrate and effector metabolites known to be present in the various plant cells (reviewed in Preiss, 1988; Preiss and Levi, 1980) are consistent with a role in starch synthesis, in contrast to the known properties of the UDPglucose specific starch synthase and plant phosphorylases. For these two enzymes the high K_m values for the respective substrates (UDPglucose and glucose-1-P) compared to the actual cellular levels preclude a significant role in starch synthesis. In view that phosphorylase catalyzes an equilibrium reaction in cells having P_i concentrations in excess of glucose-1-P it is most likely that phosphorylase plays a degradative role in starch degradation rather than synthesis. No relationship has been observed between starch synthesis and the activities of starch phosphorylase or UDPglucose pyrophosphorylase in the tissues studied.

The properties of the plant and bacterial ADPGlc PPases have been extensively reviewed (e.g., Preiss, 1988; 1991). In every plant tissue studied so far, the major activator of the ADPGlc PPase is 3-phosphoglycerate (3PGA) and the major inhibitor, P_i. In fact in a review (Preiss, 1982) a table listed ADPGlc PPases from 13 leaf sources and 9 non-photosynthetic plant sources that had been demonstrated to be activated by 3PGA. 3PGA also causes the enzyme to have better affinity for the substrates, ATP and glucose-1-P. In the non-photosynthetic tissues studied (e.g., maize endosperm, potato tuber, cassava tuber) the ADPGlc PPase activity is highly dependent on the presence of 3PGA and is inhibited by P_i. As pointed out by Hawker and Smith (1982), the ADPGlc PPase of cassava (*Manihot esculenta* Crantz) tuber ADPGlc PPase was more dependent on the presence of 3PGA for activity than the cassava leaf ADPGlc PPase. Moreover, the tuber enzyme was more sensitive to P_i inhibition than the leaf enzyme causing Hawker and Smith (1982) to state that regulation of the tuber enzyme by 3PGA and P_i was just as pronounced as the leaf enzyme and that one could not generalize that regulation of non-photosynthetic tissue ADPGlc PPase was less in magnitude than the leaf tissue enzyme.

There is some evidence suggesting that the wheat endosperm enzyme is not activated by 3PGA (M.Olive and P. Keeling, personal communication) but it is inhibited by P_i and, as observed in other plant systems, photosynthetic or non-photosynthetic, 3PGA overcomes the inhibition caused by P_i. However at least another investigator (J.S. Hawker, verbal and written communication) has seen slight activation by 3PGA, 30-to 40%, of the wheat endosperm enzyme. Moreover, J.S. Hawker has also indicated that (private communication) P_i inhibited with a Ki of 2.7 mM. Thus even in wheat endosperm regulation of starch synthesis may occur via fluctuation of the [3PGA]/[P_i] ratio. It is worth noting that partial proteolysis during enzyme isolation can strongly affect its regulatory properties. Proteolysis may be the source of some artifacts found in the literature for the maize endosperm (Dickinson and Preiss, 1969; Plaxton et. al.,1987) pea embryo (Hylton and Smith, 1992) and barley endosperm ADPGlc PPases (Kleczkowski *et al.*, 1993). The ADPGlc PPase from maize endosperm has been partially purified (88-fold, up to a specific activity of 34 μmol glucose-1-P per min per mg protein) taking great care to avoid proteolysis, using phenylmethylsulfonyl fluoride and/ or chymostatin. For the non-proteolyzed enzyme, measured in the direction of synthesis of ADPglucose, 3PGA (0.5 mM) decreased the k_m for glucose-1-P and ATP 20-fold and 7-fold, respectively, and increases V_{max} 25 to 30-fold. When proteolysis of the enzyme was not prevented, the major changes were the progressive appearance of a 53 KDa polypeptide, at the expense of the 54 KDa subunit, the increased enzymatic activity in the absence of added activator and the decrease in sensitivity to the allosteric effectors, 3PGA and P_i.

The activation of the enzyme by 3PGA and inhibition by P_i found *in vitro* seems to be physiologically important as indicated by many *in vivo* and *in situ* experiments. Many of these experiments have been cited in previous reviews (Preiss, 1982; 1988; 1991; Preiss and Levi, 1980) showing a direct correlation between the concentration of 3PGA and starch accumulation, and an inverse one between P_i concentration and starch content. This is true for photosynthetic tissues, in which P_i and PGA concentrations within the chloroplast are good indicators of the energy and carbon status, and in this way the ADPglucose pyrophosphorylase provides a good regulatory mechanism for the flux of photosynthate into starch. On the other hand, it has been found that the regulatory properties of the enzyme are also relevant to the flux of carbon in storage tissues, but it is uncertain what is its role is as a signal for the availability of carbon and energy for starch synthesis.

It is important to note recent experiments (Neuhaus and Stitt,1990; Neuhaus et al., 1989) using Kacser-Burns control analysis methods (Kacser and Burns, 1973; Kacser, 1987). This is a method where the enzyme activity can be varied (either by availability of mutants or by variation of physiological conditions) and the effect of change in enzyme activity on the possible change in rate of a metabolic process (e.g., starch synthesis) is measured. If the enzyme activity is rate limiting or is involved in the regulation of the metabolic process, then a large effect on the process should be seen. If there is no effect with change in enzyme activity then the enzyme level of activity is not considered to be rate-limiting for the metabolic process being measured. It was shown in a Kacser-Burns analysis experiment, that ADPGlc PPase in *Arabidopsis*, is a major site of regulation of starch synthesis (Neuhaus and Stitt, 1990) and that 3PGA is an important regulatory metabolite of the enzyme (Neuhaus et al., 1989). *Arabidopsis* ADPGlc PPase mutant strains containing only 50- and 7% of the normal activity, had 39% and 90% reduction in the starch synthetic rate compared to the normal plant (Neuhaus and Stitt, 1990). In experiments using a leaf cytosolic phosphoglucoseisomerase mutant of *Clarkia xantiana* having only 18% of the activity seen in the normal plant, sucrose synthetic rates were lowered but starch synthetic rates were increased (Neuhaus et al., 1989). In the mutant, the 3PGA chloroplastic levels were increased about 2-fold and Neuhaus et al. (1989) indicate that their results strongly suggest that the increase of starch synthetic rate in the cytosolic phosphoglucoseisomerase deficient mutant is due to activation of the ADPGlc PPase by the increased 3PGA concentration and 3PGA/P_i ratio.

A most significant finding indicating that 3PGA is an important physiological activator of the photosynthetic ADPGlc PPase was made by Ball *et al.*, (1991) who isolated a starch deficient mutant of *Chlamydomonas reinhardtii* which contained an ADPGlc PPase that could not be effectively activated by 3PGA.

Thus data continues to accumulate showing the importance of the plant ADPGlc PPase in the regulation of starch synthesis and that 3PGA and P_i are important allosteric effectors *in vivo*.

Starch after cellulose is the most abundant polysaccharide in nature and is the major reserve polysaccharide in green plants. About 2/3 of the total starch production is used in the food and beverage industries. It is directly applied to sauces, custards and desserts as a thickener and after enzymatic hydrolysis it is used as a sweetener in drinks and confectionery. One third of the starch production however, is used for sizing of paper and board and as an adhesive in the paper, packaging and textile industries. About 10% of the production is used as a raw material in the chemical industry for production of polyols, amino acids, acids, cyclodextrins, fructose, antibiotics and related compounds through various fermentation processes. Within the last 10 years, the demand for starch has dramatically increased for both industrial and specialized food uses (Katz, 1991). Further application of starch in the chemical industry may also result from effective competition with cellulose and crude oil, two other raw materials applied in the production of a number of the aforementioned compounds.Indeed, the use of starch relative to cellulose and crude oil may increase due to; 1) the crude oil sources are dwindling as well as its price is increasing subject to political situations and accessibility; 2) the price of cellulose fibres is more than double the price of starch and its supply is decreasing as long as pollution reduces the growth rate of the forests allover the world and 3) there is great potential in future basic and application research in that the starch granule properties can be changed or modified via molecular engineering techniques producing a starch more suitable for chemical and/or biotechnological processes. In order to improve starch for these means more investigation is needed to determine the factors affecting the properties of starch; for, e.g., the amount of protein, or lipid or covalent phosphate associated with the granule, the ratio of amylose to amylopectin, the two polysaccharide entities of the granule, as well as the structural features of these components in the granule.

RESULTS AND DISCUSSION

Increasing the levels of starch by expression in plant tissues of a bacterial ADPGlc PPase gene.

Because of the increased demand for starch it was of interest to see if starch levels in a plant could be increased by expression of activity of one of the enzymes that could be involved in starch biosynthesis. As indicated above there is a preponderance of evidence to strongly suggest that that the rate-limiting and regulatory enzyme of starch or bacterial glycogen synthesis was ADPGlc PPase (Preiss and Levi, 1980; Preiss, 1982; 1984; 1988; 1991; Neuhaus and Stitt, 1990, Ball et al., 1991). However, in view of the difficulties involved in the expression of 2 distinct genes to reconstitute the plant ADPGlc PPase activity, an *E. coli* ADPGlc PPase gene, *glg* C16, of an allosteric mutant (Leung *et al.*, 1986) that was not dependent on the presence of activator for activity, was used by the Monsanto Company group to transfect some plant systems to see if the starch content of the plant could be increased (Stark *et al.*, 1992). Since starch synthesis occurs in the plastid a nucleotide sequence encoding transit peptide of the *Arabidopsis* ribulose 1,5-bisphosphate carboxylase chloroplast transit peptide was fused to the translation initiation site of the glg C16 gene (Fig. 1). The chimeric gene was. then cloned behind the cauliflower mosaic virus (CaMV)

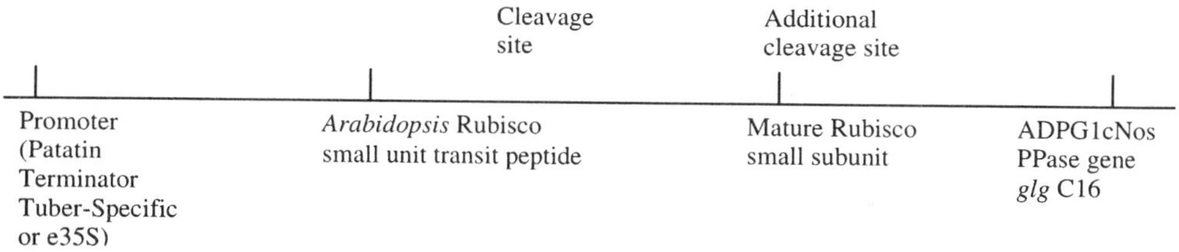

Figure 1. Construction of the synthetic promoter-plastid transit peptide-*glg* C16 ADPGlc PPase gene. The above gene was synthesized using a modified *Arabidopsis thaliana* chloroplast transit peptide portion of the Rubisco gene which also contained a duplicated cleavage site to eliminate the 23 amino acids of the mature N-terminal of the Rubisco small subunit (Stark *et al.*, 1992) so it would not interfere with the catalytic or regulatory activity of the expressed ADPGlc PPase of the *glg* C16 gene. The Nos terminator is the nopaline synthase 3′ poly A signal.

enhanced 35S promoter or a tuberspecific patatin promoter and a polyadenylation signal from the nopaline synthase gene (Nos) fused on at the 3′ end of the chimeric gene (Stark *et al.*, 1992; Fig. 1).

The CaMV-chimeric gene was placed in a cloning vector, pMON20104 having a 35Sneomycin phosphotransferase gene as a selectable marker (Stark *et al.*, 1992), mated into Agrobacterium strain ASE and used for transformation of tobacco plants (Stark *et al.*, 1992). Calli and shoots, resulting from transformation and also kanamycin resistant, were analyzed for expression of the glgC gene and starch content. The glg C16 gene product was detected in the transformed plant calli via immuno-blotting. Where the glgC gene product was detected, the starch levels were increased over the controls lacking the glgC gene product by about 1.7- to 8.7-fold in tobacco calli (Stark *et al.*, 1992). When the CaMV-chimeric gene and polyadenylation signal was subjected to electroporation into tobacco TXD protoplasts ADPGlc synthesis was observed in which was resistant to P_i inhibition and activated by fructose 1,6 bis-P (Table 1). ADPGlc synthesis in the control protoplast extract was totally inhibited by P_i as expected since the tobacco and most plant ADPGlc PPases are quite sensitive to P_i.

Examination and comparison via the light microscope of transgenic tobacco with control calli, show a tremendous increase of starch granules (Stark *et al.*, 1992). Similarly, I_2 staining of transgenic tomato shoots containing the transit peptide-Glg C16 gene caused them to stain black while the controls were essentially negative.

Table 1. The ADPGlc PPase activity in tobacco protoplasts electroporated with *glg* C16.*

Conditions	ADPglc Formed (nmol)
Control, 10mM inorganic Phosphate	0.0
Transformed, 2.5 mM Fructose 1,6 bis-P	20.2
Transformed, 2.5 mM Fructose 1,6 bis-P + 10 mM inorganic Phosphate	18.0
Transformed, 20 mM 3-Phosphoglycerate	18.4
Transformed, 10 mM inorganic Phosphate	6.4

* Data from Stark et al, (1992).

These results certainly indicate that the bacterial enzyme can be expressed in plant tissues and that it stimulates greater production of starch further suggesting that it is a rate limiting enzyme for starch synthesis.

Similar results have been obtained for Russet-Burbank potato tubers where the chimeric gene with transit peptide, the polyadenylation signal, under control of a tuber-specific patatin promoter, increases starch in the tuber, 25- to 60% over controls not containing the bacterial enzyme (Table 2; Stark *et al.*, 1992). If the bacterial ADPGlc PPase Glg C16 gene

Table 2. Starch Content in tubers expressing the *Glg* C16 gene with and without the Rubisco chloroplast transit peptide.

Conditions	No. Plant Lines	Average starch content (% wet weight)
Rubisco transit peptide-*Glg* C16	11	16.02 ± 2.0%
Control	11	12.3 ± 1.2%
Glg C16, no transit peptide	26	12.4 ± 1.0%
Control	22	12.4 ± 0.08%

was expressed in the tuber lacking the transit peptide gene portion, no increase in starch content was noted (Table 2) and this most probably is due to the expressed ADPGlc PPase activity not being being present in the amyloplast where starch is synthesized via starch synthase and branching enzyme catalysis.

What is also noted is that there is a relationship between the expression levels of the ADPGlc PPase of *Glg* C16 and the increase in starch content (Table 3). The levels of the expressed ADPGlc PPase was measured by Western blotting of the potato extracts and comparison to known standards. Starch content was measured as glucose formed after digestion with amylase and glucoamylase (Stark *et al.*, 1993). Control starch levels were obtained from 12 tubers.

A larger percent increase was noted over those tubers having lower starch content (Table 3). Lower levels of the expressed ADPGlc PPase resulted in increases of 21 to 63 %, intermediate levels of the expressed ADPGlc PPase gave increases of 33 to 118% and high expressed levels of the Rubisco transit peptide-*Glg* C16 resulted in increases of 33 to 167%.

Table 3. Effect of of expression levels of Rubisco transit peptide-*Glg* C16 on starch content*

Range of Rubisco transit peptide-*Glg* C16, Content, in tuber, ng per 50 μg of protein	Range of Starch Content, % fresh weigh
0 (control)	5.4–14.4
0.5–10	8.8–17.4
10–25	11.8–19.1
26–50	14.4–19.2

* Data from Stark et al, 1992.

These data indicate that transfection of a plant with a bacterial ADPglucose PPase will increase the level of starch in an important crop product. Thus, via biotechnological methods the increase of starch content in plants can be accomplished. It is conceivable therefore, that similar methods can be used to change in addition to quantity, starch quality via expression/transformation of the isoforms of starch synthase and branching enzymes in plants. Various isoforms of the starch synthase and branching enzyme have been isolated from various plants (Preiss and Levi, 1980; Preiss, 1988; 1991; Ozbun et al, 1972; Hawker et al, 1974; Boyer and Preiss, 1978; Singh and Preiss, 1985) and recently, the maize branching enzyme isoforms have been shown to have different catalytic properties with respect to size of oligosaccharide chains transferred (Takeda *et al.*, 1993; Guan and Preiss, 1993). It is quite possible that the starch synthase isoforms also have different properties with respect to elongation of various branch chains in amylopectin and in amylose synthesis (Preiss, 1991). Thus introduction of these various isoenzymes separately, may result in obtaining starches with modified structures in both amylopectin and amylose. These "new starches" may have greater utilization in food and industrial processes. The production of modified starches via molecular biology techniques are thus promising, may be more beneficial and even more economical than producing modified starch via chemical methods. However, in order to have a rational approach in this endeavor it would be of importance to know the various properties of the branching enzyme and starch synthase isozymes.

Branching Enzyme; Measurement of Branching Enzyme Activity

There are three assays used for the measurement of branching enzyme.

1. The iodine assay is based on the decrease in absorbance of the glucan-iodine complex resulting from the branching of amylose or amylopectin by the enzyme. During incubation of the assay mixture containing amylose or amylopectin, aliquots are taken at intervals and iodine reagent is added (Boyer and Preiss, 1978). For amylose, the decrease of absorbance is measured at 660 nm and, for amylopectin,

at 530 nm. A unit of activity is defined as decrease in absorbance of 1.0 per min. at 30° at the defined wave length.

2. The phosphorylase-stimulation assay (Boyer and Preiss, 1978; Hawker *et al.*, 1974) is based on the stimulation of the 'unprimed' phosphorylase activity of the phosphorylase *a* from rabbit muscle as the branching enzyme present in the assay mixture increases the number of non-reducing ends available to the phosphorylase for elongation. One unit of activity is defined as 1 mol of glucose transferred from glucose-1-P per min. at 30°C.
3. The branch-linkage assay (BL assay, Takeda, Guan and Preiss, 1993) is the only one to measure the number of events catalyzed by the branching enzyme, rather than an indirect effect of its action as in the two assays described above. The enzyme fraction is incubated with the substrate, $NaBH_4$-reduced amylose, the reaction is then stopped by boiling, and the product is incubated with crystalized *Pseudomonas* isoamylase for debranching. Finally, the reducing power of the oligosaccharide chains transferred by the enzyme is measured by a modification of the Park-Johnson method. Reduction of amylose with borohydride gives about 2% of the reducing power of the non-reduced amylose, resulting in lower controls. A unit of activity is defined as 1 μmol of reducing end formed after isoamylolysis per min. at 30°C.

In summary, the branching-linkage assay is the most quantitative assay for BE but amylolytic activity interferes most with this assay. The phosphorylase-stimulation assay is the most sensitive as it employs [^{14}C]glucose-1-P as a substrate. The I_2 assay is not very sensitive but enables one to test specificity with various α-1,4-dextrins. It may be best to employ all three assays to study the various properties of the BE isoforms. However, amylases can interfere with the measurement and studies of the properties of branching enzyme and, for this reason, great effort is required to completely eliminate or minimize to a large extent amylase activity in the BE preparation.

Purification and Characterization of the Branching Enzyme Isoforms

In maize endosperm there are three branching enzyme isoforms (Boyer & Preiss, 1978; Singh and Preiss, 1985; Guan and Preiss, 1993). Multiple forms of BE have been identified in other plants: pea endosperm (Smith, 1988) and rice endosperm (Nakamura *et al.*, 1992). Purification of BE I, IIa and IIb has been achieved from maize endosperm (Boyer and Preiss, 1978; Singh and Preiss, 1985; Guan and Preiss, 1993) and involves ammonium sulphate precipitation and chromatography on DEAE-Sepharose in which the three isoforms are first separated. BEI is further purified by chromatography on α-aminodecyl agarose, and by FPLC using Mono Q and Superose 12. Further purification of both BEIIa and BEIIb requires chromatography on α-aminooctyl agarose and MonoQ.

Mode of Action

Because of the difficulties in separating the isoenzymes between themselves and from other enzymes, including amylases, it is not yet clear what are the best acceptors and what are the products for each isoenzyme. Some characterization of the enzymes had been done, e.g., k_m for α-glucans such as amylopectin, different animal glycogens and maltooligosaccharides. Characterization of the products has been minimal, but progress has been made recently in our laboratory (Takeda, *et al.*, 1993, Guan and Preiss, 1993).

Takeda *et al.* (1993) have analyzed the branched products made from amylose by each BE isoform. This was done by debranching the products of each isoform using isoamylase, followed by gel filtration. BEIIa and BEIIb are very similar in their affinity for amylose and the properties of the products. When presented with amyloses of different average chain length, all the BE have higher activity with the longer chain amylose. BEI can still catalyze the branching of an amylose of average c.l. of 197 with 89% of the activity shown with the c.l. 405. The activity of BEII drops sharply with chain length. The action of BEIIa and BEIIb results in the transfer of shorter chains than those transferred by BEI. The action of the isoforms on amylopectin has been studied by Guan and Preiss (1993) and, of the three isoforms, BEI had the highest activity (using the iodine assay) in branching amylose and its rate of branching amylopectin was less than 5% of that with amylose (Table 4). In contrast, the BEIIa and IIb isoforms branched amylopectin at twice the rate they branched amylose, and catalyzed branching of amylopectin at 6-times the rate observed for BEI (Table 4). These results are consistent with the results of Takeda *et al.*, (1993) in suggesting that BEI catalyzes transfer of longer branched chains and that BEIIa and IIb catalyzes transfer of shorter chains. Thus, it is quite possible that BEI may produce slightly branched polysaccharides which serve as substrates for enzyme complexes of BE II isoforms and starch synthases to synthesize amylopectin and BE II isoforms may play a major role in forming short chains of amylopectin. In other words BEI may be more involved in formation of the larger B chains while BEII may be involved in formation of the exterior A chains and smaller B chains.

Table 4. Relative specific activities (U/mg protein) of maize endosperm BE I, BE IIa and BE IIb in the various branching enzyme assays and with amylose and amylopectin.* Units of each assay are defined in the Measurement of activity section.

Branching Enzyme	BE I	BEIIa	BEIIb
Phosphorylase stimulation	473	795	994
Branching-Linkage assay	1.04	0.32	0.14
Iodine stain assay			
amylose	320	30	39
amylopectin	9.6	59	63

*Data from Guan and Preiss, 1993.

Cloning of the Maize Branching Enzymes

The results by Guan and Preiss (1993) and by Takeda *et al.* (1993) and described above, suggest that the BE isoforms could play distinct roles in amylopectin synthesis. Recently, cDNA clones of BE have been isolated from maize (Baba *et al.*, 1991; Fisher *et al.*, 1993), pea (Bhattacharyya *et al.*, 1990), potato (Koßmann *et al.*, 1991; Poulsen and Kreiberg, 1993) and rice (Nakamura and Yamanouchi, 1992; Mizuno *et al.*, 1993). To facilitate the study of the structure–function relationships of the BE isoforms using chemical modification and site-directed mutagenesis, it is necessary to express the maize BE in *E. coli* and to obtain relatively large amounts of the purified enzyme. Therefore, we have cloned the cDNAs encoding BEI and BEII from maize endosperm and separately expressed them in *E. coli*.

A maize endosperm cDNA library (*Zea mays*, cultivar B73 endosperm, hand-dissected, 30 days after pollination) was prepared in UNI-ZAP XR (Stratagene). The library (from Monsanto Co., St. Louis) was screened with a probe produced by PCR, using oligonucleotide sequences corresponding to the segments of amino acid sequences determined after trypsin digestion of the purified maize BEII (unpublished results). Putative clones were plaquepurified and pBluescript SK-vector containing the insert was excised *in vivo* from UNI-ZAP XR following the Stratagene's protocol using helper phage R408. A clone containing the full length maize BEII cDNA inserted in the SmaI and XhoI sites of pBluescript SK, was identified by restriction mapping and sequenced by the dideoxy chain termination method.

Construction of Maize BE Expression Vectors

For expression of maize BE in *E. coli*, the maize BEI cDNA in pUC19 (Baba *et al.*, 1991) was subcloned into pBluescript SK- at the EcoRI site. The PCR method was used to modify the N-terminus of maize BE using the following oligonucleotides: primer A (5′ATACTAGTCCATG GCTACTGTGC AAGAAGATA AAAC-3′); Primer B (5′-TGTGTACCTTTCAGAAGCAGG AG-3′); Primer C (5′GATCTAGACCATGGTTC-CTGAGGGCGG AAT-3′); Primer D (5′-TACATGGTATCCAAAGCTT CC-3′). Primer A paired with primer B and Primer C paired with primer D were used to introduce a NcoI restriction site in front of the N-terminus of putative mature BEI and BEII respectively. The genes coding for mature maize BEI and BEII were separately subcloned into the NcoI/XhoI site of the bacterial expression vector pET-23d (Novagen).

Expression of Maize BE in *E. coli*

An overnight culture of the transformed cells [Bl21(DE3)] with maize BE gene was diluted 1:20 in fresh LB medium containing 100 μg/ml ampicillin and grown at 37°C for about 2 h to A_{600} = 0.6. The expression of maize SBE was induced by adding IPTG to 0.5 mM and incubating at 25°C for 12 h. Cells were harvested in a refrigerated centrifuge. Cell paste was resuspended and lysed by sonication in 50 mM Tris-acetate buffer containing 10 mM EDTA and 2.5 mM DTT. The lysed suspension was then centrifuged at 10,000 g for 15 min. BE activity was determined in the supernatant.

The genes coding for mature maize BEI and BEII were separately expressed in *E. coli* Bl23 (DE3). The specific activities for BEI and BEII obtained in the supernatant were 3.1 and 9.5 units/mg protein respectively (Table 5). SDS-PAGE followed by western blotting

Table 5. Expression of maize endosperm BEI and BEII in *Escherichia coli*. The activity of BE and protein concentration were measured in the supernatant. The amount of cells extracted was 0.5 g in a 10 mL volume. The cells with the native plasmid, pET-23d without the insert was used as the control.

Transformants	Protein conc.		
	Mg/mL	Units/mL	Units/Mg
Control, Native plasmid	1.9	0.29	0.15
Maize BEI	2.8	8.7	3.1
Maize BEII	2.4	22.8	9.5

showed that each of the expressed proteins had the same molecular mass as those displayed by the enzymes extracted from maize endosperm. Although no branching enzyme activity was found in the pellet (after centrifugation of the sonicated mixture), western blotting did show about 80% of the expressed maize BEI protein and about 40% of the maize BEII protein being present in the pellet. Antibody neutralization studies showed that antibodies raised against maize BE neutralized the expressed branching activity efficiently. In the control experiments, the antibodies raised against maize BE did not neutralize *E. coli* glycogen branching enzyme activity. The amounts of anti-serum causing 50% inhibition were about 4 and 19 μl/unit enzyme for anti-maize BEI and anti-Maize BEII respectively. These results are extremely similar to those corresponding to the enzymes purified from maize endosperm (3.4 and 18 μl/unit enzyme for anti-maize BEI and anti-maize BEII serum respectively).

In the future studies, we will take advantage of this expression system to purify large amounts of maize BE isoforms from the transformed *E. coli* instead from field-grown maize. The specific activities of the BEs (BEI and BEII) of the crude extracts from maize kernels were lower (from 2 to 3 units/mg protein, Boyer and Preiss, 1978; Guan and Preiss, 1993) than those from the bacterial expression (Table 5), and the BE isoforms from maize kernels can not be separated until the DEAE sepharose chromatography (Guan and Preiss, 1993). This expression system also makes it possible to do site-directed mutagenesis studies on the maize BE isoforms. The expression systems for maize BEI and BEII will be a useful tool to elucidate the structure–function relationships of the branching isozymes with respect to their specificity and to facilitate understanding their functions in starch synthesis.

Acknowledgements

This research was supported in Part from the United States Department of Agriculture/ Department of Energy/National Science Foundation Plant Science Center Grant 88-37271-3964 and by Michigan State University Research Excellence Funds.

REFERENCES

Baba, T., Kimura, K., Mizuno, K., Etoh, H., Ishida, Y., Shida, O., Arai Y. (1991) Sequence conservation of the catalytic regions of amylolytic enzymes in maize branching enzyme—I. *Biochem. Biophy. Res. Commun.* 181, 87–94.

Ball, S., Marianne, T., Dirick, L., Fresnoy, M., Delrue, B. and Decq, A. (1991) A *Chlamydomonas reinhardtii* low-starch mutant is defective for 3-phosphoglycerate activation and orthophosphate inhibition of ADP-glucose pyrophosphorylase. *Planta* 185, 17–26.

Bhattacharyya, M., Smith, A.M., Ellis, T.H.N., Hedley, C., Martin, C. (1990) The wrinkledseed character of pea described by Mendel is caused by a transposon-like insertion in a gene encoding starch-branching enzyme *Cell* 60, 115–122.

Boyer, C.D. and Preiss, J. (1978) Multiple forms of (1->4)-α D-glucan 6-glycosyl transferase from developing *Zea mays* L. kernels. Carbohydr. Res.61, 321–334.

Dickinson, D.B. and Preiss, J. (1969) Presence of ADP-glucose Pyrophosphorylase in *shrunken-2* and *brittle-2* mutants of maize endosperm. *Plant Physiol.* 44, 1058–1062.

Fisher, D.K., Boyer, C.D., Hannah,L.C. (1993) Starch branching enzyme II from maize endosperm. *Plant Physiol.* 102, 1045–1046.

Guan, H.P. and Preiss, J. (1993) Differentiation of the properties of the branching isozymes from maize. *Plant Physiol.* 102, 1269–1273.

Hawker, J.S., Ozbun, J.L., Ozaki, H., Greenberg, E. and Preiss, J. (1974). Interaction of spinach leaf adenosine diphosphate glucose α-1,4-glucan α-4-glucosyl transferase and α-1,4glucan, α-1,4-glucan-6-glucosyl transferase in synthesis of branched α-glucan. *Arch. Biochem. Biophys.* 160, 530–551.

Hawker, J.S. and Smith, G.M. (1982) Salt tolerance and regulation of enzymes of starch synthesis in cassava (*Manihot esculenta* Crantz) *Aust. J. Plant Physiol.* 9, 509–518.

Hylton, C. and Smith, A.M. (1992) The *rb* mutation of peas causes structural and regulatory changes in ADP-glucose pyrophosphorylase from developing embryos. *Plant Physiol.* 99, 1626–1634.

Kacsar, H. (1987) Control of metabolism. In Davies, D.D., (ed.). *The Biochemistry of Plants,* 11, Academic Press, Inc. New York, pp. 39–67.

Kacsar, H. and Burns, J.A. (1973). Control of Flux. *Symp. Soc. Exp. Biol.* 27, 65–107.

Katz, F.R. (1991) Biotechnology and Food Ingredients, Van Nostrand Reinhold, New York, p. 315–326.

Kleczkowski, L.A., Villand, P., Lüthi,E., Olsen, O.-A. and Preiss, J. (1993) Insensitivity of barley endosperm ADP-glucose pyrophosphorylase to 3-Phosphoglycerate and orthophosphate regulation. *Plant Physiol.* 101:,179–186.

Koßmann, J., Visser, R.G.F., Müller-Röber, B.T., Willmitzer, L. and Sonnewald, U. (1991) Cloning and expression analysis of a potato cDNA that encodes branching enzyme: evidence for co-expression for starch biosynthetic genes. *Molecular Gen. Genetics* 203, 237–244.

Leung, P., Lee, Y.M., Greenberg, E., Esch, K., Boylan, S., and Preiss, J. Cloning and expression of the *Escherichia coli glg* C gene from a mutant containing an ADPglucose pyrophosphorylase with altered allosteric properties. *J. Bacteriol.* 167, 82–88 (1986).

Lin, T.P., Caspar, T., Somerville, C. and Preiss, J. (1988) Isolation and characterization of a starchless mutant of *Arabidopsis thaliana* (L) Henyh lacking ADPglucose pyrophosphorylase activity. *Plant Physiology* 86, 1131–1135.

Mizuno, K., Kawasaki, T., Shimada, H., Satoh, H., Kobayashi, E., Okumura, S. Arai, Y., Baba, T. (1993) Alteration of the structural properties of starch components by the lack of an isoform of starch branching enzyme in rice seeds. *J Biol. Chem.* (in press).

Müller-Röber, B.T., Sonnewald, U. and Willmitzer, L. (1992) Inhibition of ADPglucose pyrophosphorylase in transgenic potatoes leads to sugar-storing tubers and influences tuber formation and expression of tuber storage protein genes. *EMBO J.* 11, 1229–1238.

Nakamura, Y., Takeichi, T., Kawaguchi, K., Yamanouchi, H. (1992) Purification of two forms of starch branching enzyme (Q-enzyme) from developing rice endosperm. *Physiol. Plant.* 84, 329–335.

Nakamura, Y., Yamanouchi, H. (1992) Nucleotide sequence of a cDNA encoding starch branching enzyme, or Q-enzyme I, from rice endosperm. Plant Physiol. 99, 1265–1266.

Neuhaus, H.E., and Stitt, M. (1990). Control analysis of photosynthate partitioning: Impact of reduced activity of ADPglucose pyrophosphorylase or plastid phosphoglucomutase on the fluxes to starch and sucrose in *Arabidopsis. Planta,* 182, 445–454.

Neuhaus, H.E., Kruckeberg, A.L., Feil, R. and Stitt, M. (1989). Reduced activity mutants of phosphoglucose isomerase in the cytosol and chloroplast of *Clarkia xantiana* II. Studies of the mechanisms which regulate photosynthate partitioning. *Planta* 178, 110–122.

Okita, T. (1992) Is there an alternative pathway for starch synthesis? *Plant Physiol.* 100, 560564.

Ozbun, J.L., Hawker, J.S. and Preiss, J. (1972) Soluble adenosine diphosphate glucose-α1,4-glucan α-4-glucosyltransferases from spinach leaves. *Biochem. J.* 126, 953–963.

Plaxton, W.C. and Preiss, J. (1987). Purification and Properties of Nonproteolytic Degraded ADPglucose Pyrophosphorylase from Maize Endosperm. *Plant Physiol.* 83, 105–112.

Poulsen, P., Kreiberg, J.D. (1993) Starch branching enzyme cDNA from *Solanum tuterosum. Plant Physiol* 102, 1053–1054.

Preiss, J. (1982) Regulation of the biosynthesis and degradation of starch. Annual Review of *Plant Physiology* 54, 431–454.

Preiss, J. (1988) Biosynthesis of starch and its degradation. In: The Biochemistry of plants, 14, *Academic Press,* pp. 181–254.

Preiss, J. (1991) Starch Biosynthesis and its regulation. In: *Oxford Surveys of Plant Molecular and Cell Biology,* 7, J.Miflin (ed.), pp. 59–114.

Preiss, J. and Levi, C. (1980). Starch Biosynthesis and Degradation. In: Preiss, J. (ed.). *The Biochemistry of Plants,* 3, Academic Press, Inc., pp 371–423.

Singh, B.K. and Preiss, J. (1985) Starch branching enzymes from maize: Immunological characterization using polyclonal and monoclonal antibodies. *Plant Physiol.* 78, 849–852.

Smith, A.M. (1988). Major Differences in isoforms of Starch-Branching enzyme between Developing Embryos of Round and Wrinkled-Seeded Peas (*Pisum sativum* L.). *Planta* 175: 270–279.

Smith, A.M., Bettey, M. and Bedford, I.D. (1989) Evidence that the *rb* Locus Alters the Starch Content of Developing Pea Embryos through an Effect on ADPglucose Pyrophosphorylase. *Plant Physiol.* 89, 1279–1284.

Stark, D.M., Timmerman, K.P., Barry, G.F., Preiss, J. and Kishore, G.M. (1992) Role of ADPglucose Pyrophosphorylase in regulating starch levels in plant tissues. *Science* 258, 287292.

Takeda, Y., Guan, H.P. and Preiss, J. (1993) Branching of Amylose by the branching isoenzymes of maize endosperm. *Carbohydrate Res.* 240, 253–263.

Tsai, C.Y. and Nelson, O.E. (1966) Starch-deficient Maize Mutant Lacking Adenosine Diphosphate Glucose pyrophosphorylase activity. *Science* 151 341–343.

GENETIC ENGINEERING OF RESISTANCE TO STARCH HYDROLYSIS CAUSED BY PRE-HARVEST SPROUTING

R. J. Henry,[1] G. McKinnon,[1] I. A. Haak,[2] and P. S. Brennan[2]

[1]Queensland Agricultural Biotechnology Centre
Gehrmann Laboratories
University of Queensland
Queensland 4072, Australia

[2]Queensland Wheat Research Institute
PO Box 2282
Toowoomba, Queensland 4350, Australia

SUMMARY

Pre-harvest sprouting results in significant loss of end-use quality in wheat. Several biochemical events are associated with the deterioration in quality. The production of enzymes associated with mobilisation of endosperm reserves during germination is a major factor. Several approaches to control of sprouting damage by genetic engineering have been considered. The expression of proteins capable of inhibiting the action of hydrolytic enzymes is an important option. Progress has been made towards the production of transgenic wheat with genes encoding inhibitors of the main starch degrading enzyme, alpha-amylase.

INTRODUCTION

Pre-harvest sprouting results from rain falling on mature crops and causing the grain to germinate in the field prior to harvest. This is a widespread problem occurring in many grain producing areas (Derera, 1990).

Pre-harvest sprouting results in a serious reduction in the quality of wheat (McMaster, 1987). Sprouting can cause dough handling problems and may result in a sticky crumb in bread and detract from the quality of other products produced from wheat (Henry et al., 1987).

Malting barley has a reduced storage life following sprouting (Moor, 1987). In other respects sprouting is a less serious problem in barley than it is in wheat (Ringlund, 1987).

Improvement of Cereal Quality by Genetic Engineering, Edited by
Robert J. Henry and John A. Ronalds, Plenum Press, New York, 1994

OPTIONS FOR THE CONTROL OF SPROUTING DAMAGE

Production Management

The level of sprouting damage may be limited by timing of crop growth, choice of varieties and early harvesting to ensure the grain is at risk for the minimum possible time. However, these options are very limited and are at best an option for reducing the average level of sprouting damage.

Post-harvest Management

Sprouted grain may be used for a range of purposes but the value of the grain is usually reduced significantly. Reliable analysis of grain can be used to segregate damaged grain and ensure that a small amount of sprouted grain does not reduce the value of a much larger amount of sound grain (Kruger, 1990; Henry and Blakeney, 1990). Processing of sprouted wheat to remove the damaged tissues has application in some cases (Henry et al., 1987).

Breeding Resistant Cereal Varieties

The production of cereal varieties with resistance to sprouting damage is potentially the best solution to this problem. Significant efforts have been devoted to the breeding of sprouting resistant varieties of wheat and other species (Gale et al., 1983).

GENETIC ENGINEERING OF RESISTANCE TO SPROUTING DAMAGE

Control of Germination

Genetic control of germination would appear to be a useful option for tackling the sprouting problem. Conventional genetic approaches have had limited success for several reasons. Most importantly, the genetic complexity of natural germination control processes such as dormancy have limited their exploitation in breeding. Genetic engineering to control germination requires an understanding of the molecular basis of the control of germination that is lacking at present.

Control of Alpha-amylase

Much of the loss in quality of sprouted wheat is associated with the production of alpha-amylase and other hydrolytic enzymes. Reduction in the level of alpha-amylase expression during germination could be investigated by introducing appropriate genes. For example, antisense control of alpha-amylase may be possible. However, these approaches could result in plants with reduced performance in the early stages of growth if supply of sugars from the endosperm was to become limiting. The large number of different alpha-amylase genes (Gale, 1983) could also be a problem.

Control of Alpha-amylase Inhibitors

Naturally occurring alpha-amylase inhibitors may provide an option for the control of sprouting damage (Henry et al., 1992). Barley contains a natural inhibitor of alpha-amylase, the bifunctional alpha-amylase/subtilisin inhibitor (BASI) (Mundy et al., 1983). This protein binds to and inhibits the high pI group of alpha-amylases from wheat (Battershell and Henry, 1990). Transfer of the barley amylase inhibitor gene to wheat (Henry et al., 1993) may result in some protection of wheat from sprouting damage. The gene which codes for the amylase inhibitor is located on the long arm of chromosome 2 of barley. The addition line in which the chromosome 2 of Betzes barley was added to Chinese Spring wheat was crossed with the commercial wheat cultivar Hartog (Haak *et al.*, 1993). Callus was initiated from immature embryos and regenerated plants assessed for the presence of the amylase inhibitor gene and other markers on the barley chromosome. The aim is to identify lines with the target gene that have lost most of the other barley markers during cell culture. Plants meeting these objectives have been produced and are being evaluated.

Vectors for Transformation of Wheat with the BASI Gene

As a first step towards the production of suitable vectors sequences encoding BASI have been obtained by PCR amplification of barley genomic DNA (Henry and Oono, 1991) using primers based upon the cDNA sequence (Leah and Mundy, 1989). The genomic sequence of BASI (McKinnon and Henry, unpublished) indicates the absence of introns and the presence of a 23 amino acid signal peptide. PCR tests have been developed for the detection of constructs containing the barley gene in a wheat background.

A variety of different promoters have been considered for use in this work. A vector in which BASI is placed under the control of a constitutively expressed promoter might ensure strong seed expression, but cause deleterious effects in other tissues. Hormonally controlled promoters should provide greater specificity. Vectors have been prepared with 750 bp of a barley high pI alpha-amylase promoter (Jacobsen and Close, 1991) linked in an identical manner to either BASI or GUS as a reporter gene. If correctly regulated in transgenic wheat this promoter should ensure production of BASI which is temporally linked to that of alpha-amylase. Earlier production and storage of BASI in the endosperm may be achieved by the use of an endosperm specific promoter. To this end isolation of the native BASI promoter, which we anticipate will direct a high level of expression in wheat endosperm (on the basis of the very high level of expression of BASI in barley) has commenced.

Wheat Transformation

We have established procedures for the regeneration of sprouting susceptible commercial varieties of wheat from immature embryo-derived callus and for the efficient microprojectile based transformation of this tissue at an early stage of culture. A variety of explants of the cultivar Hartog have been investigated for transformation competency and response to selective agents. Stable transformation should be possible using this approach.

Acknowledgements

This work was funded by the Grain Research and Development Corporation.

REFERENCES

Battershell, V.G. and Henry, R.J. (1990) High performance liquid chromatography of alpha-amylases from germinating wheat and complexes with the alpha-amylase inhibitor from barley. *J. Cereal Sci.*12,73–81.

Derera, N. (1990) A perspective of sprouting research. *Fifth International Symposium on Pre-harvest Sprouting in Cereals* (Ringland, K., Mosleth, E. and Mares, D.J.,Eds.) Westview Press, Boulder 3–11.

Gale, M.D. (1983) Alpha-amylase genes in wheat. *Third International Symposium on Pre-harvest Sprouting in Cereals* (Kruger, J.E. and LaBerge, D.E., Eds.) Westview Press, Boulder 105–110.

Gale, M.D., Flintham, J.E. and Mares, D.J. (1983) Applications of molecular markers in breeding for low alpha-amylase wheats. *Fifth International Symposium on Pre-harvest Sprouting in Cereals* (Ringlund, K., Mosleth, E. and Mares, D.J., Eds.) Westview Press, Boulder 167–175.

Haak, I.C., Brennan, P.S., McKinnon, G.E. and Henry, R.J. (1993) Tissue culture as a mechanism for gene transfer from barley to wheat. *Tenth Australian Plant Breeding Conference* 2, 191–192.

Henry, R.J., Battershell, V.G., Brennan, P.S. and Oono, K. (1992) Control of wheat alpha-amylase using inhibitors from cereals. *J. Sci. Food Agric.* 58, 281–284.

Henry, R.J. and Blakeney, A.B. (1990) Post harvest management of alpha-amylase. *Aspects of Applied Biology* 25, 387–393.

Henry, R. J., McKinnon, G.E., Haak, I.C. and Brennan, P.S. (1993) Use of alpha-amylase inhibitors to control sprouting. Preharvest Sprouting in Cereals 1992, (Walker-Simmons, M.K. and Ried, J.L.,Eds) *American Association of Cereal Chemists,* St Paul 232–235.

Henry, R.J., Martin, D.J. and Blakeney, A.B. (1987) Reduction in the alpha-amylase content of sprouted wheat by pearling and milling. *J. Cereal Sci.* 5, 155–166.

Henry, R.J. and Oono, K. (1991) Amplification of a GC-rich sequence from barley by a two-step polymerase chain reaction in glycerol. *Plant Mol. Biol. Rep.* 9, 139–144.

Jacobsen, J.V. and Close, T.J. (1991) Control of transient expression of chimaeric genes by gibberellic acid and abscisic acid in protoplasts prepared from mature barley aleurone layers. *Plant Mol. Biol.* 16, 713–724.

Kruger, J.E. (1990) Monitoring for sprouting damage in cereals and its implications for end-product quality. *Fifth International Symposium on Pre-harvest Sprouting in Cereals.* (Ringland, K., Mosleth, E. and Mares, D.J., Eds.) Westview Press, Boulder 321–328.

Leah, R and Mundy, J. (1989) The bifunctional alpha-amylase inhibitor of barley: nucleotide sequence and patterns of seed specific expression. *Plant Mol. Biol.* 12, 1028–1031.

McMaster, G.J. (1987) Pre-harvest sprouting in wheat—the Australian experience. *Fourth International Symposium on Pre-harvest Sprouting in Cereals* (Mares, D.J., Ed.) Westview Press, Boulder 3–14.

Moor, T. (1987) Problems encountered in storage of sprouted barley. *Fourth International Symposium on Pre-harvest Sprouting in Cereals* (Mares, D.J. Ed.) Westview Press, Boulder 212–221.

Mundy, J., Svendsen, I. and Hejgaard, J. (1983) Barley alpha-amylase/subtilisin inhibitor. 1. Isolation and characterisation. *Carlsberg Res. Comm.* 48, 81–90.

Ringlund, K. (1987) Pre-harvest sprouting in barley. Fourth International Symposium on Pre-harvest Sprouting in Cereals (Mares, D.J. Ed.) Westview Press, Boulder 15–23.

SECTION IV

IMPROVEMENT OF BARLEY QUALITY BY GENETIC ENGINEERING

POTENTIAL FOR THE IMPROVEMENT OF MALTING QUALITY OF BARLEY BY GENETIC ENGINEERING

G. B. Fincher

University of Adelaide, Waite Campus
Department of Plant Science
Glen Osmond
South Australia 5064

INTRODUCTION

Barley quality encompasses a range of chemical and physical attributes which depend on whether the grain is to be used in the preparation of malt for the brewery, as a component of stockfeed formulations, or in human nutrition. Currently, specifications for barley quality are tailored primarily for the malting and brewing industries, although agronomic properties such as yield and disease resistance are also important selection criteria in breeding programs. Thus, parameters such as grain size, dormancy, malt extract, grain protein content, development of hydrolytic enzymes for starch degradation (diastatic power), apparent attenuation levels, (1 → 3, 1 → 4)-β-glucanase potential and (1 → 3, 1 → 4)-β-glucan content represent commonly used quality indicators. Many of these quality characteristics are determined by the expression of several or many individual genes, such that the genetic manipulation of the characters by recombinant DNA technology requires a thorough understanding of the individual genes, their interactions in the expression of a particular quality characteristic, or the identification of the rate-limiting component of the characteristic. Quality factors that are determined by a single gene are more amenable to manipulation by genetic engineering and there are developing technologies through which the expression of individual genes can be either enhanced or inhibited, depending on the desired effect on malting quality. In this paper, the potential for engineering barley to enhance (1 → 3, 1 →4)-β-glucanase activity is explored, together with the biochemical, genetic and physiological information that necessarily underpins genetic engineering technology.

Improvement of Cereal Quality by Genetic Engineering, Edited by
Robert J. Henry and John A. Ronalds, Plenum Press, New York, 1994

DEFINITION OF DESIRABLE CHARACTERISTICS

The first problem that must be addressed in genetic engineering is to define the characteristic to be altered in precise biochemical terms. For example, maltsters and brewers might wish to enhance diastase activity through genetic engineering, but the target gene or genes in such an exercise are not immediately apparent. Diastase is a measure of the starch-degrading capacity of malt extracts and includes the activities of a-amylases, β-amylases, limit dextrinases and glucosidases. Which is the rate-limiting enzyme? If it were shown to be, say, limit dextrinase, we would then need to define the number of limit dextrinase genes in barley and which of these genes is the most important for starch degradation in the germinating grain. Most of the enzymes that participate in endosperm mobilization during germination are encoded by small gene families, but not all of the genes are expressed in the grain. Thus, definition of the desirable characteristic might require extensive preliminary studies on the physiology and biochemistry of malting enzymes.

In the case of the (1 → 3, 1 → 4)-β-glucanases, enzyme levels have been closely correlated with malt extract (Stuart *et al.*, 1988) and it has been clearly demonstrated that there are only two (1 → 3,1 → 4)-β-glucanase isoenzymes in barley (Woodward and Fincher, 1982a; Slakeski *et al.,* 1990). The (1 → 3,1 → 4)-β-glucanase system is therefore relatively simple from both a biochemical and genetic viewpoint, and represents a good target for genetic engineering. Enhanced enzyme development during germination would be expected to go hand-in-hand with higher extract values, and retention of high (1 → 3, 1 → 4)-β-glucanase levels in malt would help overcome problems attributable to (1 → 3, 1 → 4)-β-glucans in the brewing process. The most serious of these problems relates to the propensity of (1 → 3, 1 →4)-β-glucans to form aqueous solutions of high viscosity; these highly viscous solutions slow down wort separation and many significantly retard the rate of beer filtration. Furthermore, (1 - 3, 1 - 4)-β-glucans that are carried through into the final product may precipitate out of solution, particularly at low temperatures and high ethanol concentrations, and thereby contribute to undesirable haze formation in the beer.

ISOLATION OF THE TARGET GENE

Having identified the (1 → 3, 1 → 4)-β-glucanase genes as targets for genetic engineering, the technology for their isolation and characterization is relatively straightforward. Genomic DNA libraries for barley are commercially available and the preparation of appropriate cDNA libraries is now routine in well-equipped laboratories. Indeed, full-length cDNAs encoding the barley (1 → 3, 1 → 4)-β-glucanases isoenzymes EI and EII have been isolated and sequenced (Slakeski *et al.,* 1990), while the corresponding genomic DNA clones are also available (Litts *et al.,* 1990; Slakeski *et al.*, 1990; Wolf, 1991).

MANIPULATION OF THE ISOLATED TARGET GENE

Once the target gene has been isolated and characterized, how might it be manipulated *in vitro* to enhance expression levels? In the case of the (1 → 3, 1 → 4)-β-glucanase gene, as with other genes important in malting quality, there are several approaches that might be taken to increase expression levels. Firstly, extra copies of the gene might be inserted into the barley genome in the expectation that multiple copies of the gene might be expressed where only one copy was originally expressed, and that this would lead to elevated levels of the enzyme. While this is a relatively simple approach, it is by no means certain that it would work. Indeed, there are many instances where additional gene copies result in decreased expression, through the cosuppression phenomenon.

A second approach could be to create a chimeric gene, in which expression is driven by an appropriate, but more powerful promoter. Thus, the high pI barley a-amylase gene promoter might be spliced onto the coding region of the (1 → 3, 1 → 4)-β-glucanase isoenzyme EII gene. The α-amylase promoter would be expected to direct gene expression in the aleurone layer during germination, as required, but levels of the (1 → 3, 1 → 4)-β-glucanase enzyme would be raised by the action of the more powerful a-amylase promoter. This approach to increasing (1 → 3, 1 → 4)-β-glucanase levels is perfectly feasible, although care must be exercised in the selection of the promoter. If the promoter were too powerful, it might direct very high levels of (1 → 3, 1 → 4)-β-glucanase synthesis at the expense of other critical enzymes in the germinating grain. Aleurone layers have a finite store of metabolic energy and must use this stored energy to synthesize a range of hydrolytic enzymes. If too much energy is directed to the synthesis of a single enzyme, other enzymes would soon become limiting and the energy balance of the cell would be upset.

A third approach to increasing (1 → 3, 1 → 4)-β-glucanase levels is to improve the stability of the enzyme itself. Both isoenzymes are very unstable at temperatures above 45°C, either in the purified form or in extracts of grain (Woodward and Fincher, 1982a; Loi *et al.,* 1987). More than 60% of enzyme activity is lost during commercial kilning protocols and any remaining activity is quickly lost during mashing (Loi *et al.*, 1987). If the stability of the enzymes could be enhanced through genetic engineering, (1 → 3, 1 → 4)-β-glucanase in green malt could survive through the kilning and mashing phases, and the enzyme would be able to hydrolyse residual (1 → 3, 1 → 4)-β-glucan in the mash. If the original (1 → 3, 1 →4)-β-glucanase promoter were retained, this approach would not be expected to perturb the metabolic balance of aleurone or scutellar cells during germination.

Once a decision is taken to genetically engineer an enzyme to enhance thermostability, the question arises as to what manipulations will be made. The enzyme would be engineered at the DNA level, using either the cDNA or genomic clones. The technology for specific, site-directed mutagenesis is well developed and kits for specifically altering gene sequences are commercially available. Indeed, Doan and Fincher (1992) engineered a small improvement in thermostability in barley (1 → 3, 1 → 4)-β-glucanase isoenzyme EI through the addition of an N-glycosylation site into the protein. However, the theoretical basis for protein stability is not always well-defined, despite the volume of work that has been directed to the problem. Intramolecular interactions such as ion bridges, disulphide cross-linking, hydrogen bonding and hydrophobic interactions will all contribute to protein stability, but it is not always possible to predict the results of changes in these interactions. Before embarking on such an exercise, it is essential to know the precise three-dimensional conformation of the protein that is to be engineered. This requires crystallization of the enzyme and the solution of the x-ray diffraction pattern. Crystallization of proteins is often difficult and the solution of structure requires sophisticated x-ray diffraction data collection facilities, high levels of expertise in the discipline of x-ray crystallography and enormous computing power. The barley (1 → 3,1 → 4)-β-glucanase isoenzyme EII has recently been crystallized (Chen *et al.,* 1993) and its three-dimensional structure determined by x-ray crystallography at 2.2–2.3 Å resolution. This opens the way for rational protein design, in which intramolecular stabilization could well lead to enhanced thermostability of the enzyme.

TRANSFORMATION OF BARLEY

A major technological bottleneck in the improvement of malting barleys by genetic engineering has been our inability to transform barley. Barley immature embryo and microspore cultures can be readily established and maintained, foreign DNA can be introduced into the cells using the microparticle gun, direct transfer into protoplasts, or electroporation, and transient expression of the foreign DNA has been monitored. Until very recently, however, the stable integration of foreign DNA into the barley genome and the regeneration of fertile, transgenic barley plants has not been achieved. Recent reports (Lemaux *et al.*, 1993) provide good evidence for the transformation of barley and, subject to the successful adoption of the technology in other laboratories, we can now be confident that this major impediment to the genetic engineering of barley has been overcome. The new challenge will be to identify useful genes or to design modifications to improve the performance of existing genes, in order to improve the malting quality of barley. As mentioned above, this requires a detailed knowledge of genes that determine malting quality characteristics.

REFERENCES

Chen, L., Garrett, T.J.P., Varghese, J.N., Fincher, G.B. and Høj, P.B. (1993) Crystallization and preliminary X-ray analysis of (1,3)- and (1 → 3;1 → 4)-β-D-glucanases from germinating barley. *J. Mol. Biol.*, in the press.

Doan, D.N.P. and Fincher, G.B. (1992) Differences in the thermostabilities of barley (1 → 3,1 →4)-β-glucanases are only partly determined by N-glycosylation. *FEBS Letters,* 309, 265-271.

Lemaux, P., Wan, Y and Williams, R. (1993) Development and Use of Transformation Systems for Maize & Barley for Agronomic Improvement. *Chem. in Aust.*, 60, 496.

Litts, J.C., Simmons, C.R., Karrer, E.E., Huang, N. and Rodriguez, R.L. (1990) The isolation and characterization of a barley (1 → 3, 1 → 4)-β-glucanase gene. *Eur. J. Biochem.* 194, 831-838.

Loi, L., Barton, P.A. and Fincher, G.B. (1987) Survival of barley (1 → 3,1 → 4)-β-glucanase isoenzymes during kilning and mashing. *J. Cereal Sci.* 5, 45-50.

Slakeski, N., Baulcombe, D.C., Devos, K.M., Doan, D.N.P. and Fincher, G.B. (1990) Structure and tissue-specific regulation of genes encoding barley (1 → 3,1 → 4)-β-glucan endohydrolases. *Mol. Gen. Genet.* 224, 437-449.

Stuart, I.M., Loi, L. and Fincher, G.B. (1988) Varietal and environmental variations in (1 → 3, 1 →4)-β-glucan levels and (1 → 3, 1 → 4)-β-glucanase potential in barley: relationships to malting quality. *J. Cereal Sci.* 7, 61-71.

Wolf, N. (1991) Complete nucleotide sequence of a *Hordeum vulgare* gene encoding ((1 → 3, 1 →4)-β-glucanase isoenzyme II. *Plant Physiol.* 96, 1382-1384.

Woodward, J.R. and Fincher, G.B. (1982) Purification and chemical properties of two (1,3:1,4)-β-endohydrolases from germinating barley. *Eur. J. Biochem.* 121, 663-669.

GENETIC MODIFICATION OF BARLEY FOR END USE QUALITY

Sietske Hoekstra, Marion van Zijderveld,* Sandra van Bergen, Frits van der Mark, and Freek Heidekamp

Center for Phytotechnology RUL/TNO
Department of Molecular Plant Biotechnology
Wassenaarseweg 64, 2333
AL Leiden
The Netherlands

SUMMARY

Since the late 70s both recombinant DNA technology and plant cell and tissue culture techniques have been applied to change the properties of crop plants. Ten years later, these technologies have been used more or less routinely to modify specific traits in a large number of dicotyledoneous plants, including important crop plants like tomato, potato, cotton and oil seed rape. The application of plant cell and tissue culture techniques as well as DNA-transfer technology to cereal crops has made considerable progress in the past decade. This has resulted in stable transformation methods for [hybrid] maize, rice and wheat in the early 90s.

In 1989 we initiated a research program aiming at the stable transformation of barley. To achieve this, the availability of efficient procedures to regenerate barley plants from cells and/or tissues is essential.

In this paper an overview is presented of our work on the development of an efficient and reproducible method of plant regeneration from barley microspores. The current protocol enables us to regenerate 50–100 green fertile barley plants from a single anther. The use of microspore cultures in studies on cell differentiation will be described. Finally, the use of transformation technologies to improve malting quality of barley will be discussed.

* Address any correspondence to Dr. van Zijderveld.

Improvement of Cereal Quality by Genetic Engineering, Edited by
Robert J. Henry and John A. Ronalds, Plenum Press, New York, 1994

KEY FACTORS FOR EFFICIENT AND REPRODUCIBLE MICROSPORE CULTURE IN BARLEY

Application of anther and microspore culture in breeding programs has so far been limited due to a lack of an efficient and reproducible microspore culture system.

Such a system requires the consistent production of a minimum number of (doubled) haploid plants from a given number of anthers. The definition of the exact numbers will depend on the crop, and is reflected in the level of homozygosity and the inbreeding depression. Nevertheless, it is always necessary that embryo-like structures (ELS) develop into plants (conditions concerning turnover of embryo's into plants will not be dealt with here). ELS are derived from dividing microspores, which in this section will be defined as the embryogenic microspore type. The ultimate goal is to be able to distinguish embryogenic microspores in a mixed population of embryogenic and non-embryogenic microspores.

Upon isolation three different microspore-subpopulations can be distinguished by visual selection (Hoekstra et al, 1993). They can be recognised by 1) the diameter, 2) the colour interference of the exine, and 3) the colour and appearance of the cytoplasm. The smallest subpopulation is non-viable. The remaining two subpopulations are viable, but from these, the fraction characterised by the smaller diameter, will be plasmolysed after 3 days in culture. These two viable microspore types can further be distinguished by the colour of the exine. It should be mentioned that the colour of a viable microspore can vary, depending on the physiological condition and the medium applied. In our conditions, the plasmolysing fraction will show a blue, and the embryogenic subpopulation will show a red colour interference of the exine. Moreover, the colour of the exine and the microspore diameter are influenced by the osmotic environment (Hoekstra et al, unpublished). Besides the osmotic pressure other components influence the colour of the exine and the diameter of the microspore. Olsen (1991) stated that embryogenic microspores can be distinguished by a blue colour interference of the exine. From a comparison of the two methods it was concluded that in these experiments a lower osmotical value of the culture medium as well as a different microscope was used.

For initiation of microspore divisions, a minimum density of embryogenic microspores is required. One explanation for this observation maybe that viable microspores secrete substances which are crucial for the formation of ELS. In the shed microspore system, where microspores become isolated from the anther without any mechanical forces, a limitation in the plating density is essential. When microspores form an ELS within the anther however, there is no need for a minimum plating density. During culture the breakdown of anther wall and tapetum provide nutrients to the microspores. It is known, that the anther wall is a rich source of e.g. glutamine and proline. Concerning the plating density of a culture, the presence of non-viable microspores should be noticed, as they release detrimental substances which inhibit the development of embryogenic microspores. For example, a population which consists of 95% non-viable microspores requires a 10 times higher total density compared to a population where 50% of the microspores are viable. The production of a culture where only 5% of the microspores are of the embryogenic type will never be of the same magnitude.

The negative influence of the non-viable population can be compensated for by adjusting the density of embryogenic microspores. In summary, the composition of the population determines the efficiency of the culture. With the above consideration, microspore culture can show reproducible efficiencies in development, independent of the pretreatment or isolation method applied.

The composition of the isolated microspore population can be highly influenced by the isolation method. Sofar, two methods of isolation have been described in the literature. Swanson et al (1987) and Olsen (1991) described improvement of microspore isolation using a Waring blender compared to the conventional method of mechanical isolation using a teflon rod, a Potter homogenizer or a syringe. The ideal method largely depends on the degree of damage caused to the tissue. When microspores are isolated by squashing, the applied pressure should be very gentle. Nevertheless, in our laboratory the best results have been obtained by squashing anthers with a teflon rod: routinely, about 60% of the microspores isolated are of the embryogenic type, compared to a maximum of 30% by blender isolation. Using the mechanical isolation method, there is no need for further purification of the isolated microspores. It should be mentioned, that the microspores are only cultured when at least 30% of the population is of the embryogenic type. On the contrary, when microspores are isolated by blending, a purification step using, e.g., Percoll improves the culture efficiency. However, a lower number of microspores will be obtained due to this extra step in the isolation procedure.

The blending method can be advantageous when extraordinary large amounts of microspores are needed, e.g., in transformation experiments. With the blending method complete spikes can be processed. Under these circumstances, more than 20 spikes per day can be used for the isolation of microspores. Two prerequisites should be mentioned, namely pretreatment of the whole spike and abundant donor material. Huang and Sunderland (1982) reported the cold pretreatment of whole spikes by storage of the material at 4°C for 4 weeks. The material should be reselected after the incubation period, as it can become either desiccated or drowned, even in a well-controlled climate chamber. An alternative pretreatment, the so called starvation method, has been established by Roberts-Oehlschlager (1990) where anthers are harvested and incubated for 3 to 5 days on a mannitol solution. In our laboratory the anther starvation is routinely applied for 4 days. Compared to the cold pretreatment the following advantages of the starvation method have been encountered: 1. no reselection is necessary as the anthers show a reproducible homogeneous response (also with other cultivars) after the short incubation period, 2. danger of infection is reduced as the source for microorganisms is removed, 3. contamination is detected during the incubation as mannitol provides redundant nutrients for active growth of the majority of microorganisms.

The duration of pretreatment depends on the developmental stage of the microspore population. In general, the microspore developmental stage of a spike is determined for the anthers of the middle floret. When the cell cycle stage of the microspores is early, a longer pretreatment is necessary. For example, a miduninucleate stage requires 4–5 weeks of cold pretreatment and a late uninucleate stage 3–4 weeks; for anther starvation, the differences in time will be at the level of hours. In our hands the optimal stage is when 50% of the microspores are in the midlate-to-late uninucleate stage, with less than 5% in the bicellular stage and less than 20% in the mid uninucleate stage (Hoekstra et al, 1992). The window of the responsive stage is very small, and has a direct relation to the homogeneity of a population and the optimum plating density.

The homogeneity of the microspore population is influenced by the selection of the starting material and the growth conditions of the donor plants. The selection of the starting material is not limited to the stage test of the middle floret per spike. Also the number of florets per spike and the number of spikes per plant used are important. There is a non-homogeneous development in a spike, and this developmental pattern varies per genotype. In some genotypes the youngest florets are located at the top of the spike, whereas in other genotypes these can also be found at the bottom of a spike. The representativity of the stage test of the middle flower will be highly affected by the pattern of floret development within a spike.

Besides, the spike number of a plant has an impact on the homogeneity of the obtained microspore population. In general, the first spike is a little more progressed in development compared to the next 5 spikes. At a higher spike number, more variation in spike morphology and microspore development have been observed. Therefore selection of a higher spike number for microspore isolation will show less productivity, unless a thorough selection has been made.

The selection of the material, including the stage determination, is highly affected by the growth conditions of the donor plants. Under low temperature and high light intensity conditions, there will be a slow, more synchronised development, which increases the chance of harvesting material in the appropriate microspore developmental stage. Besides, the plants are more uniform in morphology, compared to greenhouse or field material and this increases the chance to distinguish the spike in the right stage. Experience with the selection of material derived from well-controlled growth conditions, enables recognition of the appropriate material from field or greenhouse-grown plants which will develop into doubled haploid plants at reasonable frequencies.

It should be emphasized that only 25% of the so called embryogenic subpopulation does divide after 3 days in culture. This implies that on average 15% of the total microspore population is destined to develop into a plant. Research is in progress to further characterise this specific microspore population.

Anther and microspore culture provide a tool for hastened selection of improved genotypes as they are uninucleate single cells with a haploid genome. With this method a new wheat variety called Florin has been produced (de Buyser et al, 1987). Furthermore, the application of in vitro selection or mutagenesis will aid the breeding process. Recently, with the combination of microspore culture and particle bombardment, transgenic barley plants have been obtained (pers. comm. Prof. dr. H. Lörz, Hamburg, Germany). With this important result the use of microspores will enable directed augmentation of genetic variability of barley in the future.

BARLEY CHARACTERISTICS WHICH MAY BE IMPORTANT FOR IMPROVED END-USE QUALITY

In parallel with the availability of efficient transformation and regeneration procedures for food crops the question can be addressed which economical important traits should be modified in order to improve the quality and use of these crops. For barley the following traits can be considered : 1) resistance to diseases and plague insects, 2) improved food and feed quality by increasing the nutritional value and 3) improved malting quality. In our group the metabolism of lipids in barley and their importance for malting quality is being studied. Lipids may be important for a number of reasons (Kokkelink et al, 1992). They may play a role in foam stability, in fermentation by yeast and may cause filtration problems in the brewhouse. In addition, degradation products of lipids may play a role in beer staling or ageing. As was suggested by Drost and co-workers and others (Drost et al,1990; Martel et al, 1993) oxidation of lipids by lipoxygenase may lead to the formation of off-flavours like trans-2-nonenal in the end-product. Therefore it was decided to study the role of lipoxygenase in this process. As a first step, lipoxygenase was purified from green malt (Doderer et al, 1992). Two protein preparations were isolated, lox 1 and lox 2 which were used for the generation of monoclonal antibodies. Lox 1 and lox 2 have a MW of approximately 90 kD, have the same IEP but yield different products. Lox 1 converts linoleic acid into the corresponding 9-hydroperoxide whereas lox 2 solely forms the 13-hydroperoxide (Doderer et al, 1992, see

Figure 1). Because trans-2-nonenal is a breakdown product of the 9-hydroperoxide of linoleic acid this may suggest that lox 1 is the enzyme which is responsible for the formation of this off-flavour and thus for staling of the beer.

Further research will be conducted to elucidate the role of lipoxygenases in the degradation of lipids during germination of barley.

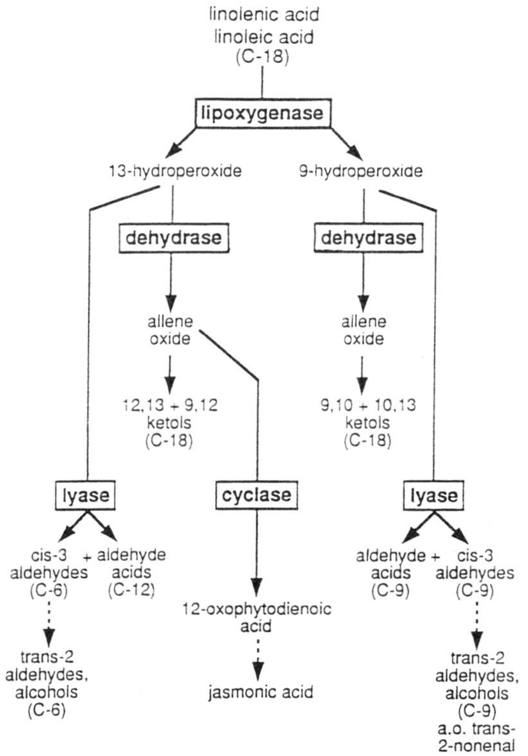

Figure 1. The lipoxygenase pathway in plants resulting a.o. in the formation of off flavour compounds (Doderer et al., 1991).

Acknowledgements

The authors would like to thank Jan Vink for expert technical assistance. This work was supported by EUREKA grant nr. EU270 and is ABIN publication 130.

REFERENCES

Doderer A., I. Kokkelink, S. van der Veen, B.E. Valk, A.C. Douma. Purification and characterization of lipoxygenase from germinating barley. *EBC Congress 1991;* Chapter II, pp. 109–116.

Doderer A., I. Kokkelink, S.W. van der Veen, B.E. Valk, A.W. Schram and A.C. Douma. Purification and characterization of two lipoxygenase isoenzymes from germinating barley. *Biochemica et Biophysica Acta* 1120 (1992) 97–104.

Drost BW, Van den Berg R, Freijee FJM, Van der Velde EG and Hollemans M: Flavor stability. *Am. Soc. Brew. Chem.* 48: 124–131 (1990).

Hoekstra S., M.H. van Zijderveld, J.D. Louwerse, F. Heidekamp and F. van der Mark. Anther and microspore culture of *Hordeum vulgare L.* cv. *Igri Plant Science* 86 (1992) 89–96.

Hoekstra S., M.H. van Zijderveld, F. Heidekamp and F. van der Mark Microspore culture of *Hordeum Vulgare L.*: the influence of density and osmolarity. *Plant Cell Reports* 12 (1993) 661–665.

Huang B. and Sutherland N. Temperate stress pretreatment in barley anther culture *Ann. Bot.* 49 (1982) 77–88.

Kokkelink I., A. Doderer, B.E. Valk and A.C. Douma. Lipid content and -composition of quiescent barley grain cv Triumph *Proc. of the Tenth Int. Symp. on Plant Lipids,* Jerba, Tunesia, April 27–May 2 (1992) 220–223.

Martel C, Kohl S and Boivin P. Importance de la lipoxygenase sur la formation des composes carbonyles au cours du brassage. *Louvain Brewing Letters* 7th year 1: 13–23 (1993).

Olsen F. Lok : Isolation and cultivation of embryogenic microspores from barley (*Hordeum vulgare L.*). *Herditas* 115, 255–266 (1991).

Roberts-Oehlschlager SL and Dunwell JM: barley anther culture: pretreatment on mannitol stimulates production of microspore-derived embryos. *Plant Cell Tiss. Org. Cult.* 20, 235–240 (1990).

Swanson EB, Coumans MP, Wu SC, Barsby TL and Beverdorf WD: Efficient isolation of microspores and the production of microspore-derived embryos from *Brassica napus. Plant Cell Rep.* 6, 94–97 (1987).

SECTION V

REGULATION OF CEREAL GENETIC ENGINEERING

THE REGULATION OF THE USE OF GENETICALLY ENGINEERED CEREALS AS FOODS

Simon Brooke-Taylor, Clive Morris, and Carolyn Smith

National Food Authority
PO Box 7186
Canberra MC, ACT 2610
Australia

INTRODUCTION

Before discussing the specifics of genetically engineered foods, I would like to give an overview of regulation as applied to all foods. Food legislation in Australia is generally considered to fall within the legislative responsibilities of the States and Territories and each State and Territory Government is responsible for food sold within its boundaries.

To facilitate uniformity, States and Territories have, therefore, variously implemented a set of uniform legislation, the NHMRC Model Food Act, into their State/Territory legislation. Despite its name, the Model Food Act is not legislation in its own right but sets out a blueprint for States and Territories to introduce uniform food legislation within their own jurisdictions and administrative structures

Of particular note, the Model Food Act prescribes offences in connection with the preparation, packaging and sale of food which is unfit for human consumption or food which

State and Territory legislation implementing the NHMRC Model Food Act

NSW	Food Act 1989
Victoria	Food Act 1984
Queensland	Food Act 1981
Western Australia	Health Act 1911
South Australia	Food Act 1985
Tasmania	Public Health Act 1962
Northern Territory	Food Act 1986
Australian Capitol Territory	Food Act 1992

Improvement of Cereal Quality by Genetic Engineering, Edited by
Robert J. Henry and John A. Ronalds, Plenum Press, New York, 1994

is adulterated, as well as prohibiting false and misleading claims and the sale of food which is not of the nature, substance or quality demanded by the purchaser.

THE AUSTRALIAN FOOD STANDARDS CODE

Food standards implemented under the State food legislation are set out in the Australian Food Standards Code. All food sold in Australia, whether domestically produced or imported must, therefore, comply with the Code. The Code primarily standardises foods in terms of food categories rather than recipes. Where food is not covered by a standard it may still be sold but is considered to be a non-standardised food. It can only be included in a standardised food where "other foods" are specifically permitted by the Code.

In 1991, a Commonwealth and State/Territory agreement on uniform food standards laid the foundation for a single national statutory authority to set uniform national food standards. The outcome of this agreement was the National Food Authority (NFA) which was established by the *National Food Authority Act* in 1991. Standards developed by the NFA require approval by a majority of Commonwealth and State/Territory Health Ministers but are thereafter adopted by reference, and without variation, into all appropriate State & Territory food legislation. Furthermore, States and Territories have agreed not to develop individual food standards outside of the agreement.

THE NATIONAL FOOD AUTHORITY

As a statutory authority, the NFA fulfils a number of specific functions laid down in the NFA Act.

The Functions of the National Food Authority

- The Development of Standards
- Co-ordination of Food Recalls
- Research And Surveys
- Co-ordination of Food Surveillance
- Food Safety Education
- Development of Assessment Policies in Relation to Imported Food
- Development of Codes of Practice for Industry
- Advice to Minister (Commonwealth)
- Incidental Functions

THE OBJECTIVES OF FOOD STANDARDS

A criticism of the standards setting system that preceded the NFA was that it was not accountable and did not have clearly definable objectives. In order to provide a clear and agreed basis for the development of food standards, the NFA Act prescribes 5 objectives to which the Authority must have regard.

The Objectives of Food Standards (in priority order).

- The protection of public health and safety
- The provision of adequate information relating to food to enable consumers to make informed choices and to prevent fraud and deception
- The promotion of fair trading in food
- the promotion of trade and commerce in the food industry
- the promotion of consistency between domestic and international food standards where these are at variance, providing it does not lower the Australian standard

The breadth of these objectives are probably unique amongst national food regulatory agencies, particularly in respect of 4 and 5. And they present us with a rare opportunity to build an innovative regulatory system which actively promotes and supports the internationally competitive position of Australian food.

REGULATION IN RESPECT OF GENETICALLY ENGINEERED CEREALS

Turning now to genetic engineering, it is appropriate to note that although it is this technology which has provided the major regulatory focus to date and is, therefore, where many examples are drawn from, in regulatory terms it is only one facet of the wider field of modern biotechnology.

FOODS AND FOOD INGREDIENTS PRODUCED BY GENETIC ENGINEERING

In considering the food regulator's role in respect of genetic engineering it is important to emphasise that the focus is on the food offered for consumption, not upon the means of production. That is the responsibility of bodies such as the Genetic Manipulation Advisory Committee (GMAC) and its proposed successor the Genetic Technology Approval Authority (GTAA). Nonetheless, the NFA in attempting to identify possible outcomes will need to have regard to relevant technology and therefore seeks to integrate its processes with those of these bodies.

Foods and food ingredients produced from biotechnology cover an enormous range, from simple to complex products, from those with a well established history of safe consumption to those with which we have very limited experience. In order to facilitate a rational approach to assessing, not only safety but also consumer information and fair trade issues, it is practical to separate these foods and food products into manageable groups.

Foods and Food Additives

Food products from genetically engineered foods can be divided readily into foods, and food additives, a division based upon the physical, chemical and physiological characteristics which have already led to differences in regulatory treatment for their traditional counterparts..

Foods do not normally require individual safety evaluation and listing in the Food Standards Code before they may be sold. Certain plants which are consumed as food are acutely toxic when raw and require particular processing before they are eaten, for example

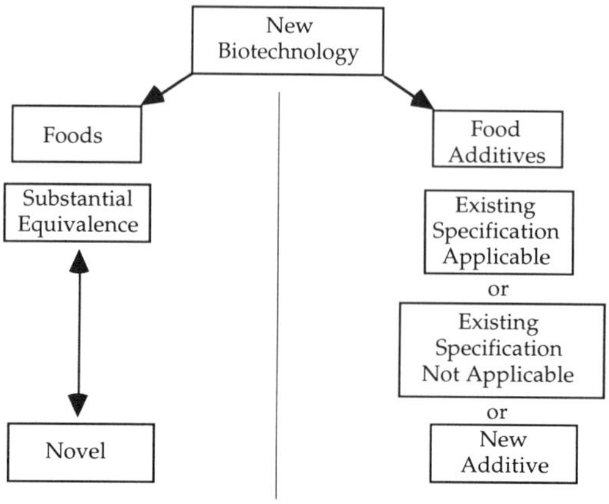

Figure 1. Categorisation of foods and food ingredients produced by biotechnology.

red kidney beans. We have traditionally relied upon historical knowledge and experience for our safety evaluation. In the few cases where a new plant variety has proved to contain an unacceptable level of a natural toxin it has rapidly disappeared from agricultural use.

Beer and bread were two of the earliest foods to be standardised—for definitional purposes, to prevent consumer fraud and deception, to maintain the nutritional quality of staples and to facilitate trade.

More recently a number of new varieties derived from plants which are not normally considered suitable for human consumption have been standardised, for example canola and sweet lupins. In both cases, a maximum permitted concentration for toxins has been established. However, given the existing provisions of the State Food legislation, this action probably has more effect in the provision of confidence with respect to these new varieties than it does with the protection of public safety.

In contrast to foods, we require the safety of food additives (and processing aids), whether from natural or synthetic sources, to be established before they are approved for use. In order to be approved they are subjected to detailed safety evaluation and their use must be shown to be technologically necessary.

NOVEL FOODS AND THE CONCEPT OF SUBSTANTIAL EQUIVALENCE

GMO products can also be divided on the basis of whether they are substantially equivalent to foods which have a well established history of safe consumption, or whether they are novel and do not have a counterpart with a history of food consumption.

For example, a herbicide resistant cereal variety produced by a intra-species genetic manipulation introduces nothing different from that which be expected from a conventional cross bred variety. From a regulatory perspective, there would be a requirement the establishment of a Maximum Residue Limit (MRL) for the herbicide in that cereal but, otherwise, the product is substantially equivalent to its traditional counterpart. Is it not reasonable, therefore, to treat such a food as normal?

What if the variety has been produced from species which do not interbreed but which are both widely consumed as food? One might consider whether the donor organism has undesirable characteristics normally inactivated by processing which is not appropriate to the host organism? One might also attempt to assess what effects the expression of the inserted gene might have on the biochemistry and physiology of the host organism. If we are talking about wheat, will it, for example, still have the same nutritional value?

Now consider that the donor organism is known to produce allergenic responses in some consumers. What is the potential for the new organism to display allergenic activity characteristics of the donor organism? Such effects rely on the new organism producing proteins characteristic of the donor organism. The technology used to select and insert the target gene can be very specific. Would it not be highly unlikely that the genes coding for the allergenic characteristics would also be transferred?

Another possible manipulation is the insertion of a viral gene to produce a virus resistant variety. Would the presence of viral coat protein make the new variety different. True, viral coat protein has not previously been consciously added to food but it has been present, as a structural component of the viral coat, in virus infected grain. Can we therefore conclude that although the new variety is transgenic it is substantially equivalent to conventional grain and should not require extensive premarket testing.

The value of the substantial equivalence approach for whole foods is that where divergences are identified, assessment, whether in regard to safety, labelling or other criteria, can be focussed on those differences. Where there is no basis for comparison with a traditional counterpart then the food or food ingredient must be assessed on the basis of its own composition and properties .

The same principles can be applied to food additives produced from genetically engineered organisms. Each food additive has a specification, a clear and usually unambiguous description of the product. The majority of specifications do not describe the technology used to produce the substance. The specification for modified starch, for example, does not specify the cereal source, let alone the variety. So, in the majority of cases it is relatively straightforward to determine whether an additive is novel.

There is however potential for confusion where a specification is applicable more by omission than by intent. For example, in a number of cases, specifications will have been prepared, on the basis of an established production process, at a time when modern biotechnology was not anticipated. While the active ingredient (the food additive) from a new process or organism may be chemically identical, other criteria which are not specified, for example new impurities, may differ. Consideration must, therefore, be given to the intended scope of any existing specification before applying the substantial equivalence approach.

LABELLING

Traditionally, food products are labelled according to their content and food manufacturers are not required to indicate on the label the processes by which the food is made. Whether contents labelling is sufficient for foods produced using new biotechnology is one of the most complex issues facing the Authority. As a general principle, labels should provide information which is both useful and meaningful and which enables consumers to make an informed choice and to prevent fraud and deception.

One may debate whether without considerable educational input, advice on the technology used will assist consumers in making informed choices about food. Labelling *per se* will not necessarily meet this objective but can product marketing?

It is also important in regard to labelling to ensure that any requirements are consistent with those of our major trading partners and can be enforced. To require labelling of all products of GMOs for example could rapidly lead to a situation where domestic producers were penalised because their operations could be audited while imported foods flouted the regulations without fear of prosecution.

CONCLUSION

The Authority has to date considered it prudent, when faced with new biotechnology, to take a cautious approach in the first instance. It has prepared a proposal to develop a standard for GMOs which will enable it to assess novel products but will also permit a generic approach to be taken for those products which are no longer considered novel. It has also published a discussion paper on the safety assessment and labelling of genetically manipulated organisms.

In reaching this position the Authority noted that the specific regulation of GMOs may be perceived to be unfair when compared with conventionally bred organisms. However, it is also aware that the specific approval of the suitability of GMOs for food use, by a national regulatory agency, could be desirable from a marketing perspective. Such affirmation by an independent, national regulatory body may also be an important tool in the promotion of high quality, safe Australian food for export. This is one of the roles which the Authority can undertake to support innovation in the Australian food industry.

Finally, we have provided a lot of questions and few answers, because I believe you, the producers, know the answers best. The debate on the use of genetic engineering is a hot issue in food regulation at the present time. It is essential that all aspects of the technology, including the continuity of biotechnology, are addressed and that undue attention is not focussed narrowly on the exotic transgenics.

REFERENCES

"Final Report of the Policy review" NFA, Canberra May 1993.

"Genetic Manipulation: The Threat Or The Glory?", Report by the House of Representatives Standing Committee on Industry, *Science and Technology,* February 1992.

"Strategies for Assessing the Safety of Foods produced by Biotechnology", *Report of a Joint FAO/WHO Consultation,* World Health Organisation Geneva 1991.

"Safety Evaluation of Foods derived by Modern Biotechnology - Concepts and principles", *OECD* (Organisation for Economic Co-operation and Development), Paris 1993.

"The Labelling of Foods Sourced from Genetically Modified Organisms: Revised Guidelines." *Food Advisory Committee Consultation Paper* , UKMAFF, April 1993.

"Assessing the safety and consumer information requirements of genetically manipulated organisms in food", *NFA,* July 1993.

RAPID CEREAL GENOTYPE ANALYSIS

H. L. Ko and R. J. Henry

Queensland Department of Primary Industries
Queensland Agricultural Biotechnology Centre
Gehrmann Laboratories
The University of Queensland
Queensland 4072, Australia

ABSTRACT

Identification of cereal species and varieties is potentially useful in cereal production, storage, handling and processing. Labelling of transgenic foods may require the availability of reliable testing methods. Genotype analysis has been simplified by the development of methods based upon the polymerase chain reaction (PCR). Adaptation of this technique to the routine analysis of cereals is being investigated, with the objective of developing a reliable, simple and rapid protocol. Sampling, sample preparation and analysis procedures have been evaluated for several species. These studies have included an evaluation of PCR performance of DNA extraction methods and PCR cycles in different formats. Oligonucleotide primers with both arbitrary and specific sequences have proven successful in cereal analysis. Random amplification of polymorphic DNA (RAPD) can be applied in comparing and distinguishing varieties, while specific primers are required for species identification.

INTRODUCTION

In commercial grain trading (handling, marketing, processing) and in cereal breeding, it is of great importance for cultivars with particular commercial value to be distinguished from one another, whereas in products containing a mixture of cereals identification of the species may present a problem. Distinctions are not always possible using phenotypic characteristics of the grain, and, moreover, in many instances reliable and objective confirmation of cultivar or species identity is required (Ko *et al.*, 1994). This will become of even greater importance with the development of transgenic cereals differing by only one or a few genes.

Biochemical markers are currently the most frequent method of ascertaining the identity of cereals. There is a need to develop techniques which offer potentially simple,

Improvement of Cereal Quality by Genetic Engineering, Edited by
Robert J. Henry and John A. Ronalds, Plenum Press, New York, 1994

rapid and reliable methods for routine cereal analysis and identification. The development of polymerase chain reaction (PCR) techniques in the late 1980s has dramatically simplified genotype analysis of biological samples (Saiki *et al.*, 1988).

The application of PCR methods to grain identification offers many advantages over earlier approaches, these are:

1. The greater sensitivity of PCR, allowing the analysis of very small samples
2. The simplicity of much of the procedure, which allows automation, indicating a potential for very rapid results at low cost.

A general PCR method has the potential to distinguish all grain samples using a standard protocol (Ko *et al.*, 1994).

MATERIALS AND METHODS

Grain samples of barley, maize, oat, rice, rye, sorghum and wheat were obtained from several centres around Australia.

The following factors were addressed in this research:

1. Genomic DNA extraction:

- Several methods were being investigated to determine the simplest one, giving good quality DNA suitable for PCR.
- Performance of DNA extracted from leaves and grain were compared for efficient amplification in PCR. DNA extraction directly from grain is important if rapid results are required (no delay to allow for germination) or when the seed is very hard or impossible to germinate.

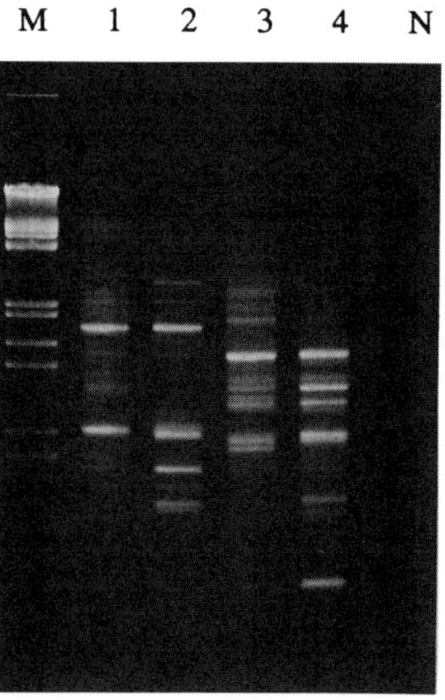

Figure 1. PCR products obtained from cereal DNA with arbitrary primers. Primer: 5′-CAAACGTCGG-3′ (OPA-19) was used to distinguish oat cultivar 'Winjardie' (*lane 1:* DNA extracted from leaves - L, *lane 2*: DNA extracted from grain - G) from barley cultivar 'Grimmett' (*lane 3*: L and *lane 4:* G). A negative control (no DNA template) is indicated in *lane N* and DNA size markers Lambda/Eco RI/HindIII (Progen Industries Ltd) in *lane M*.

Genomic DNA was extracted from grain and 1–2 week- old germinated seedlings in hexadecyltrimethylammonium bromide (CTAB) buffer (Graham *et al.*, 1994; Henry and Oono, 1991) and the concentration in the sample determined from the absorbance at 260 nm.

2. PCR:

- *Primers.* Polymerase chain reactions were carried out with random amplified polymorphic DNA (RAPD) 10-mer primers at an annealing temperature of 36°–37°C, 5S ribosomal RNA primers at an annealing temperature of 56°C (Ko *et al.*,1994) and α-amylase primers at an annealing temperature of 55°C.

- *Format.* Conditions for PCR on the thermal cyclers (Perkin Elmer 480 in tubes, PE 9600 in plates) and capillary system (Corbett) are optimised, to ensure reproducibility of the results.

3. Analysis:

- PCR products are currently analysed on 1.5% (w/v) agarose gels and stained with ethidium bromide. Techniques for colorimetric analysis for immediate assessment of presence or absence of amplification product(s) will be developed.

RESULTS AND CONCLUSIONS

- Grain DNA was found to amplify comparable main, strong bands to leaf DNA for PCR with RAPD primers, although grain DNA appears to amplify smaller products (Fig 1).

- Selected RAPD primers are suitable for comparisons between samples and in some cases for identification of cultivars if sufficient polymorphism can be found (Fig 2),

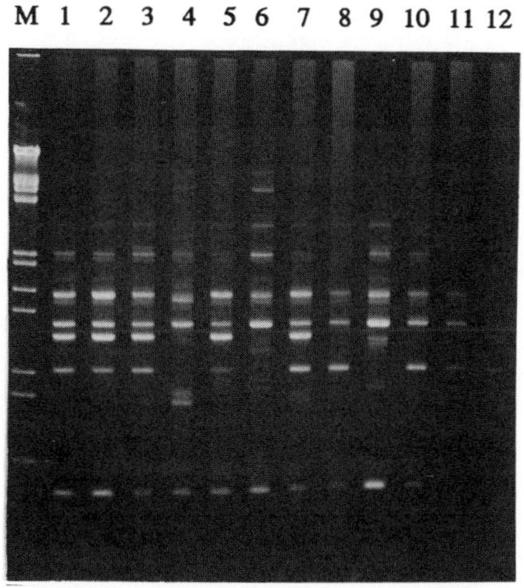

Figure 2. Comparison of PCR products from 12 rice cultivars obtained with an arbitrary primer 5′-CCACACTACC-3′ (OPJ-13). The following rice cultivars were used: Pelde (*lane 1*), YR 71003-9 (*lane 2*), CI 9187 (*lane 3*), IR 579 (*lane 4*), Bogan (*lane* 5), Purple Rose (*lane 6*), Tarra 140 (*lane 7*), Early Caloro (*lane 8*), I-geo-tze (*lane 9*), Caloro (*lane 10*), Caloro II (*lane 11*), Kulu (*lane* 12). DNA size markers Lambda/ Eco RI/ Hind III (Progen Industries Ltd.) are indicated in *lane M.*

provided consistency exists between individual plants within a cultivar. These primers are not suitable for species identification or component identification in mixed samples.

- 5S ribosomal RNA primers are particularly suitable for species identification, as they amplify a more conserved region of the gene than RAPD primers do (Table 1). These amplification fragments may be used as species specific markers. However, 5S ribosomal RNA primers are not suitable for cultivar identification, as there is no intra-specific variation, nor is it suitable for component identification in mixed samples.

Table 1. Species specific markers obtained after PCR amplification with 5S ribosomal RNA primers

Species	Marker sizes (bp)
Barley	290
Maize	260
Oat	235
Rice	240
Rye	380
Sorghum	230 and 485
Wheat	270

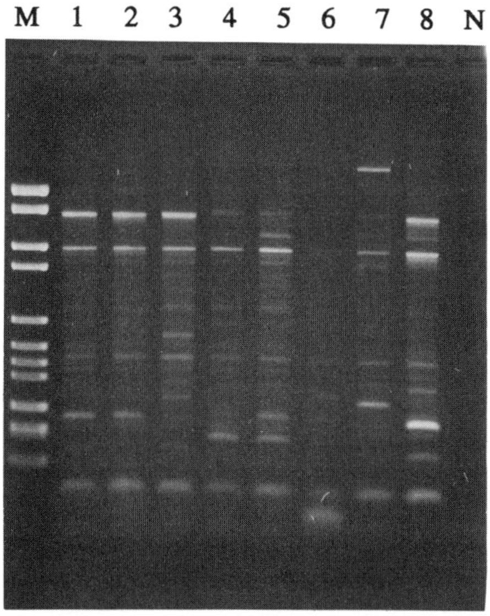

Figure 3. PCR products from 8 sorghum cultivars: SA 281 (*lane 1*), QL 27 (*lane 2*), QL 41 (*lane 3*), Kulum (*lane 4*), QL 12 (*lane 5*), SC 35C (*lane 6*), IS 12736 (*lane 7*), QL 39 (*lane 8*) , amplified with α-amylase primers 5′-CCGCCGTCGCACTCCGTC-3′ and 5′-CACCTTGCCGTCGATCTC-3′. DNA size markers VI (Boehringer Mannheim Biochemica) are indicated in *lane M* and a negative control in *lane N*.

- Selected α-amylase primers will in some cases express sufficient polymorphism to distinguish cultivars (Fig 3; Ko and Henry, 1994), and under certain conditions they are suitable for species identification in mixed samples (Weining *et al.*, 1994).

In general it can be concluded that for reliable identification of species in a mixed sample, specific oligonucleotide sequences may be developed from the amplified products of PCR with 5S ribosomal RNA and α-amylase primers.

Selected RAPD and α-amylase primers are suitable for distinguishing a range of cultivars from one another.

It has to be stressed that conditions under which the reactions were performed should be consistent to be able to reproduce the results reliably.

ACKNOWLEDGEMENT

This research was funded by the Grain Research and Development Corporation.

REFERENCES

Graham, G.C., Mayers, P., and Henry, R.J. (1994) A simplified method for the preparation of fungal genomic DNA for PCR and RAPD analysis. *BioTechniques* 16: 48–50.

Henry, R.J., and Oono, K. (1991) Amplification of a GC-rich sequence from barley by a two-step polymerase chain reaction in glycerol. *Plant Molecular Biology Reporter* 9:139–144.

Ko, H.L., Henry, R.J., Graham, G.C., Fox, G.P., Chadbone, D.A., and Haak, I.C. (1994) Identification of cereals using the polymerase chain reaction. *Journal of Cereal Science* (in press).

Ko, H.L.and Henry, R.J. (1994) Identification of barley varieties using the polymerase chain reaction. *Journal of the Institute of Brewing* 19, 101–106.

Saiki, R.K., Gelfand, D.H., Stoffel, S., Scharf, S.J., Higuchi, R., Horn, G.T., Mullis, K.B. and Erlich, H.A. (1988) Primer-directed enzymatic amplification of DNA with a thermostable DNA polymerase. *Science* 239, 487–491.

Weining, S., Ko, H.L. and Henry R.J. (1994) Polymorphisms in α-*amy1* gene of wild and cultivated barley revealed by the polymerase chain reaction. *Theoretical and Applied Genetics* (in press).

Williams, J.G.K., Kubelik, A.R., Livak, K.J., Rafalski, J.A., and Tingey, S.V. (1990) DNA polymorphisms amplified by arbitrary primers are useful as genetic markers. *Nucleic Acid Research* 18, 6531–6535.

PROSPECTS FOR GENETIC ENGINEERING IN THE OVERALL CONTEXT OF CEREAL CHEMISTRY RESEARCH*

C. Wrigley

CSIRO Grain Quality Research Laboratory
North Ryde, New South Wales, 2113
Australia

An essential prerequisite of attempting improvement in cereal-grain quality by genetic engineering is a knowledge of the chemistry and genetics of grain quality. We must know what genes to target in our manipulation studies. Preferably, we need to have isolated and characterised the genes, knowing what quality attributes they control (or at least are hypothesised to be associated with).

One hundred years of cereal chemistry research has brought us much of this essential information. The year 1993 has been suggested (Blakeney and Wrigley, 1993) as being the year for celebrating the centenary of cereal chemistry, based partly on the date of initiation of research on wheat chemistry and quality by Frederick Bickell Guthrie in Australia, but also on the basis of parallel studies in other countries.

This chapter aims to provide a historical setting for our plans in embarking on the second century of cereal chemistry research, examining the wider range of potential (including genetic manipulation), but also looking ahead for possible difficulties—"pot-holes" in the road ahead.

POT-HOLE NUMBER ONE—ACCEPTING POPULAR OPINION

Our 1993 conference contrasts markedly with the only possible equivalent held just over one hundred years ago—that is the first Rust-in-Wheat Conference convened in 1890 to deal with the question of combating rust which was greatly reducing the yield of wheat in the Colony of New South Wales (McLean et al., 1890). Farrer, the Australian pioneer of wheat breeding, contributed an anonymous letter to this conference reminding the agricultural com-

* Based on the presentation "Planning for the second century of cereal chemistry; potential, possibilities and pot-holes", at the 43rd Australian Cereal Chemistry Conference.

Improvement of Cereal Quality by Genetic Engineering, Edited by
Robert J. Henry and John A. Ronalds, Plenum Press, New York, 1994

munity even at that stage of the need to be aware of the quality of the grain as well as its yield (Farrer, 1898). This concern of Farrer's for grain quality led him to approach the Department of Agriculture in the colony of NSW, leading in turn to the research effort of Frederick Bickell Guthrie starting probably in the latter half of 1893 (Blakeney and Wrigley, 1993). Presumably Farrer, and later Guthrie, saw a connection between rust in wheat and its quality, because Guthrie reported in 1894:

"The analysis of wheats was instituted, in the first instance, with the object of determining the amount of gluten in the different varieties of wheat, the gluten content indicating the milling value of the grain, and being an important factor in the choice of wheats in connection with their ability to resist rust".

Here, perhaps, we have the first of the "pot-holes" for us to avoid as we move into the second century of cereal chemistry, that is, making unreasonable assumptions without evidence even though they may be popular opinion at the time.

LABOURING UNDER BASIC MISCONCEPTIONS LAST CENTURY

Related to this is the general difficulty, under which Guthrie laboured, of inadequate knowledge at that time about protein chemistry. The term "protein" was suggested as far back as 1838 by Berzelius in a letter to Mulder (Hartley, 1951). The term was used to describe the "organic oxide being prepared from a range of plant and animal sources, gluten being a major member of this family". The major misapprehension of the time was, however, that protein was one compound, uniform in all its sources except for the number of protein radicals and their associations with sulfur and phosphorus. According to this scheme, gluten and casein were $Pr_{10}S_2$ and ovalbumin was $Pr_{10}S_2P$, the protein radical in all of these (Pr) being the same, irrespective of whether the protein came from animal or plant origin. A further remarkable assertion was that the protein radical was absorbed as a whole unit in the digestive system and incorporated as this unit into the human body. Finally there is the assertion last century also that all plants contain only one protein known as "glutin" (Wrigley, 1993).

In the midst of these misconceptions, it was remarkable that Osborne and Vorhees could publish in 1893 details of the demonstration that wheat gluten, washed from dough, could be fractionated into gliadin (extractable into 70% aqueous ethanol) and glutenin (requiring dilute alkali or acid for dissolution) (Osborne and Vorhees, 1893). This report presumably quickened the interest of Guthrie in the protein chemistry of wheat, and he attempted to explain variations in wheat quality in terms of the proportion of gliadin. Here then we come to another "pot-hole" in that he and other chemists appear to have been particularly careless in the conditions of their extraction and quantitation of gliadin, so that the proportions of gliadin found in the flour by Guthrie and reported by others from England, France and America varied from 22% up to 80%, these variations being generally attributable in hindsight to differences in the procedures used.

THE "CAUSE" OF DOUGH STRENGTH—LAST CENTURY AND NOW

Guthrie's early enthusiasm for the use of gliadin content as an indication of dough strength was reported on the basis of a very small number of experiments and samples, without replication and with no attempt at statistical analysis. Nevertheless, he reported these results on as few as ten samples in three separate publications during the 1890's, on each occasion asserting the validity of gliadin content as an explanation of dough strength (Guthrie, 1896). Another "pot-hole" to avoid: the lack of proper statistical examination of results. Some years later, Guthrie had to write (1912), in relation to his further experimentation on the gliadin:glutenin ratio:

"Further work on this subject has convinced me that the relationship is not as simple as I had first thought; nor is the separation and accurate determination of the two proteins quite satisfactory. This method, has, therefore, been abandoned in this laboratory, and is not, I believe, any longer recognised. What is the cause of strength—still remains to be solved?"

Some might cynically agree with Guthrie, commenting that the question of dough strength still remains to be solved even today, but we can certainly claim that we know a lot more about the "separation and accurate determination of the two proteins", namely gliadin and glutenin. So today, 100 years on from these early beginnings, we can claim not only to know about the many polypeptides that make up gliadin and glutenin and thus the glutenin complex, but we can also fill many of the gaps in the sequence from the genes that code for these proteins through to the various attributes that go together to constitute wheat quality, on through to the consumers' assessment of quality.

DIVERSIFICATION OF WHEAT USES

In this picture of ours, we can speak not only of wheat quality with respect to pan-bread production, but we also know today a lot about the requirements for a much wider range of uses of wheat around the world. This market pull has changed from an accent on bread wheats needed in the "western" wheat-growing countries to a more recent accent on the wider range of products such as flat-breads for Middle East countries, noodles for the nearby Asian markets and Chinese steamed breads (Wrigley, 1994). To satisfy these diverse quality requirements, the extension of the Australian wheat-segregation process to provide up to 45 "market-based grades" has been predicted (Condon, 1993).

That range of processing uses is likely to go beyond the current range of grain-based products for world trade. In particular, wheat technologists may find that they have to cater for "foodaceutical" demands, namely, that good food is seen to have special medical benefits. Of course, the cereal grains are well placed for this role, attested to by their prominence in nutritional guidelines. However, a Japanese company has gone much further with a non-allergic rice, from which the globulin protein has been removed—being the source of allergen for many who have developed allergy to rice (Swinbanks and O'Brien, 1993). The other new role for cereal grains "waiting in the wings" is the use of cereal starches in biodegradable thermoplastics, a potentially high-value and high-volume use given the pushes from environmental lobbies for "compostable" products and for a switch from petroleum feed stocks (Roper et al, 1993).

NEW APPROACHES TO BREEDING FOR GRAIN QUALITY

What are the prospects for us being able to produce wheats to satisfy the markets of the next century of cereal chemistry? The answer to this question starts with the breeder and the ability of cereal chemists to help in the efficient selection of wheats suited to those markets. Will we see an increase in haploid breeding, a new approach that would permit the early generation selection on the basis of quality for single plants without the complicating factor of heterozygosity (Luckett and Darvey, 1992)?

The introduction of such techniques as haploid breeding brings with it new opportunities for quality segregation in breeding, particularly using techniques such as image analysis and whole-grain analysis by near infrared techniques, such that there is no destruction of the grain and the small amounts of each line are still available for growing if they are not rejected in the process of screening. These breeding techniques also accentuate the value of very small-scale testing methods, such as we have been pursuing with the 2g Mixograph. There is thus also a need for micro milling such as would suit single plants or single heads, leading in turn to the production of small flour samples for dough mixing and product testing.

Also relevant will be the possibility of new gene sources coming in to broaden the scope of wheat breeding, particularly new sources of genes from alien, ancestral wheats and wheat relatives, the identification of valuable existing but unused genes (possibly those for tolerance to environmental factors such as heat stress). However, we need to be careful of the "pot-holes" that we have struck in the past whereby alien genes (such as from rye) have brought with them problems with respect to dough quality.

Obviously there are new possibilities and great potential for genetic manipulation and the whole range of DNA technologies. These will first show their value in screening methods where we can check virtually any part of a plant for the appropriate genes with DNA-probing techniques. Next come the possibilities of accelerated breeding techniques where a specific gene (for example for hardness or softness) might be inserted by transformation in a matter of weeks instead of the necessity for repeated back-crossing over many years.

The greatest potential of genetic manipulation involves the introduction of distant genes from completely different organisms, as is happening at the moment with genes for pest resistance from bacteria being inserted into crops such as potatoes. In addition, there are the exciting possibilities of engineering genes and inserting these. For example, as we learn more about the chemical nature of gluten, we are in a position to engineer and create "super-gluten" genes that might be as effective at a lower protein content as are the natural ones. In this case, we are not too far off the goal of producing these genes as is indicated in this volume by chapters such as those by Anderson and Bekes.

These exciting possibilities prompted Dr Rafalski to describe the future laboratory in these words at a recent Plant Breeders' Conference:

"Extrapolating from currently existing technology, it is possible to envisage the plant breeding environment of the year 2000. In addition to currently available phenotypic assays, biochemical and genetic analysis results will be directly available to the breeder to facilitate selection. The breeding station of the future will incorporate a highly automated laboratory for instrumental analysis of protein, carbohydrate and oil composition, and for the determination of the allelic composition of individuals at multiple loci. The resulting data will be stored in a data base available on-line to each breeder, through a highly intuitive graphical interface." (Rafalski et al 1993).

But in all the excitement we must also watch out for the "pot-holes" in these possibilities as has already been indicated with various transgenic plants (Kareiva, 1993). To anticipate the difficulties in acceptance of new genotypes produced from genetic manipulation, CSIRO and some collaborators have organised a display to inform the public of genetic manipulation and this has been touring the various shopping centres of Australia (Alexander, 1992).

GRAIN-QUALITY TESTING

New testing technologies will also be a feature of the next century of cereal chemistry. I have already mentioned the possibility of whole-grain analysis by image analysis or near infrared methods in relation to breeding, and these methods will be equally applicable to quality checking right on through the sequence from harvest to processing. We have already seen the value of antibodies produced against specific proteins or other compounds to give us very specific and quantitative information about their presence, particularly with respect to the quality characteristics of wheat, the gluten content of foods, and the presence of pesticides in grain and grain products (Skerritt, 1991). This family of techniques has even been adapted to the detection and quantitation of insects in grain (Bair and Kitto, 1993). Some of these testing methods already provide either great specificity for insect species or general indication of insect infestation, together with speed and convenience in the testing procedure. Only a step or two ahead of antibody testing is the use of biosensors, which have the capability of continuous monitoring of whatever analyte they have been designed to determine. In this case the sensing mechanism may be an antibody or it may be one of many other sensors that can be linked electronically to provide a readout of analyte.

EXPERT SYSTEMS

The potential for on-line analysis that is offered by these new testing technologies will be important to more efficient processing and the possibility of using expert systems to increase efficiency. Such "decision-support" systems have already proved their value in agriculture where the farmer has the possibility of better management of planting time, agronomy and harvesting. We can turn them to advantage with respect to cereal chemistry in using them to predict crop quality both in variety trials for breeding applications and in relation to the harvest as a whole. In fact, there are already systems becoming available to determine what will be the average protein content and yield of cereal crops to be harvested at a certain location given the history of crops in that location and the climatic and soil conditions of the past season. Increased efficiency in storage and handling also are suited to the expert-system approach, an example being the newly completed Pestman program of the CSIRO Division of Entomology in Canberra and our own Safestor program for predicting "safe" storage times and conditions for malting barley (Bason et al, 1993). Increased efficiency in milling, baking and other grain processing is also possible by the development of expert systems working on the same principle, but a major difficulty in developing these systems is the lack of knowledge of how the various grain-processing methods operate. It seems that milling is a good place to start, given its increasing complexity—over 80 different flours in a wide range of packaging produced currently by Goodman Fielder (Breden, 1993) and the possibility of on-line quantification of bran particles in wheat flour (Evers, 1993).

OTHER NEW TECHNOLOGIES

Might we also see in the future, completely different approaches to grain transport and processing? For example, pipeline slurry technology, where grain can be transported long distances in a slurry of liquid, if a suitable medium can be found. Liquid carbon dioxide is one possibility that has been trialed. Others involve rape-seed oil; yet such technologies are still seen to be a fair way off from reality. Other new processing techniques involve genetically engineered micro-organisms some of which might be able to provide a solution to the current difficulty of omitting "chemical additives". For example, could we imagine a yeast producing bromate or other agents that would act in the same way as compounds that are now no longer accepted additives? Already it is possible for us to produce gluten proteins in a heterologous system, involving yeast or bacteria (such as *E. coli.*). It is thus quite possible that gluten might be produced in a tissue-culture system. It is more likely that a form of "super gluten" might be produced having properties different and better from native gluten, yet being produced in quantities that are economically feasible.

Finally, what might the laboratory and research environment of the coming century be? Presumably there would be increased possibilities for international collaborations, already facilitated by fax and e-mail. It might also be a competitive environment with the potential "pot-holes" of less open exchange of intellectual property and more limited interaction. Perhaps too, we will see more amalgamation of laboratories and the setting up of specialist groups. But in turn, if this is likely to happen, it will require greater interaction between laboratories if it is going to be successful.

CONCLUSION

Have we been able to learn anything from the past century of cereal chemistry? I have suggested some of the "pot-holes" into which Guthrie fell and suggested a few that we might avoid as we press on to explore the potential and possibilities of the coming century. I have already forgotten that one of these "pot-holes" is attempting to predict the future. Only ten years ago, I was invited to do so at the 33rd Australian Cereal Chemistry Conference in Brisbane (Wrigley 1983). On that occasion I said:

"The ultimate contribution that the grain-quality researcher could make to the grain industry would be to identify the genes for specific aspects of quality and to devise the means of transferring the appropriate genes from one to another. Then we could say to the breeder: 'Breed for yield and we will insert the genes for quality when you are ready'". (Wrigley, 1983). Although we have come a long way with identifying the "genes for specific aspects of quality", it still remains for us to be able to "insert genes" routinely into wheat, let alone solve all the breeders' worries about quality in the process.

REFERENCES

Alexander, N. (1992) Will pigs fly? *Search* 23:210–211.

Bair, J., and Kitto, G.B. (1993) New methods for rapidly detecting insect problems. *World Grain.* 11 (3):13–17.

Bason, M.L., Ronalds, J.A., and Wrigley C.W. (1993) Prediction of 'safe' storage life for sound and weather-damaged malting barley. *Cereal Foods World* 38:361–363.

Breden, P. (1993) Trends in wheat processing. Outlook '93 Canberra. Pp 1–4. Aust. Bureau Agric. Resource Economics, Canberra.

Blakeney, A.B., and Wrigley, C.W. (1993) 100 years of cereal chemistry in Australia. *Chem. Aust.* 60:459–460.

Condon, C.E. (1993) Meeting specific market requirements for wheat. Outlook '93 Pp 1–5. Aust. Bureau Agric. Resource Economics, Canberra.

Evers, A. (1993) On-line quantification of bran particles in white flour. *Food Sci. Technol. Today* 7: 23–27.

Farrer, W. (1898) The making and improvement of wheats for Australian conditions. *Agric. Gazette NSW* 9:131–168, 241–260.

Guthrie, F.B. (1912) Wheat and flour investigations. Sci. Bulletin No. 7. NSW Dept of Agric., Sydney.

Guthrie, F.B. (1896) The absorption of water by the gluten of different wheats. *Agric. Gazette NSW.* 7:583–589.

Hartley, H. (1951) Origin of the word "protein". *Nature* 168:244.

Kareiva, P. (1993) Transgenic plants on trial. *Nature* 363:580–581.

Luckett. D.J., and Darvey, N.L. (1992) Utilisation of microspore culture in wheat and barley improvement. *Aust. J.Bot.* 40: 807–828.

McLean, P., Pearson, A.N., Lawrie, W., Shelton, P.M., and Anderson, H.C.L. (1890) Rust in Wheat. Report of Committee appointed to draw up a series of resolutions. *Agric. Gazette of NSW* 1:41–43.

Osborne, T.B., and Vorhees, C.G. (1893) The proteids of the wheat kernel. *J. Amer. Chem. Soc.* 15:392–471.

Roper, H., Koch., and Bahr, K.H. (1993) Developments in the use of starch in biodegradable thermoplastics. *Agro-Food-Industry Hi-Tech* 4(2):17–18.

Rafalski, J.A., Hanafey, M.K., Tingey, S.V., and Williams, J.G.K. (1993) In: "Focused plant improvement, Vol. 1" (Eds. B.C. Imrie and J.B. Hacker) Proc. 10th Aust. Plant Breeding Conf. Pp 229–232.

Skerritt, J.H. (1991) Applications of ELISA technology in cereal genetic screening, production and processing. *Agro Industry Hi-Tech.* 2:29–37.

Swinbanks, D., and O'Brien, J. (1993) Japan explores the boundary between food and medicine. *Nature* 364:180.

Wrigley, C.W. (1983) Research on quality. Proc. 33rd Australian Cereal Chemistry Conf. Pp 20. Royal Aust. Chem. Instit., Melbourne.

Wrigley, C.W. (1993) A molecular picture of wheat quality: finding and fitting the jigsaw pieces. *Cereal Foods World* 38:68–74.

Wrigley, C.W. (1994). Developing better strategies to improve grain quality for wheat. *Aust. J. Agric. Research* 45: 1–17.

CONTRIBUTORS

O Anderson
USDA Western Regional Research Centre
800 Buchanan Street
Albany, CA
USA

H Anzai
Meiji Seika Ltd.
Pharmaceutical Research Centre
Japan

R Appels
CSIRO
Division of Plant Industry
GPO Box 1600
Canberra ACT 2601
Australia

G Barry
Plant Sciences Technology
The Agricultural Group
The Monsanto Company
St Louis MO 63198
USA

F Bekes
CSIRO
Division of Plant Industry
Grain Quality Research Laboratory
North Ryde NSW 2113
Australia

S van Bergen
Center for Phytotechnology
RUL/TNO
Department of Molecular Plant Biotechnology
Wassenaarseweg 64
2333 AL Leiden
The Netherlands

A E Blechl
USDA Western Regional Research Centre
800 Buchanan Street
Albany CA
USA

PS Brennan
Queensland Wheat Research Institute
P O Box 2282
Toowoomba Qld 4350
Australia

R Brettell
CSIRO
Division of Plant Industry
GPO Box 1600
Canberra ACT 2601
Australia

S Brooke-Taylor
National Food Authority
P O Box 7186
Canberra MC ACT 2610
Australia

A Castillo
Laboratory of Plant Cell and Molecular Biology
1143 Fifield Hall
University of Florida
Gainsville, FL 32611-0692
USA

R N Chibbar
Plant Biotechnology Institute
National Research Council
110 Gymnasium Place
Saskatoon
Saskatchewan
Canada S7N 0W9

R Chikwamba
ENDA-Zimbabwe
P O Box 3492
Harare
Zimbabwe

A H Christensen
University of California
Berkeley/U.S.D.A.
Plant Genome Expression Centre
USA

J Davies
Department of Botany
University of Durham
Science Laboratories
South Road
Durham DH1 3LE
UK

E S Dennis
CSIRO
Division of Plant Industry
GPO Box 1600
Canberra ACT 2601
Australia

S K Dhir
Agriculture Group of Monsanto
700 Chesterfield Parkway North
St Louis MO 63198
USA

R. D'Ovido
Department of Agrobiology and Agrochemistry
University of Tuscia
Viterbo
Italy

G B Fincher
University of Adelaide, Waite Campus
Department of Plant Science
Glen OSmond SA 5064
Australia

M E Fromm
Monsanto Company
700 Chesterfield Village Parkway
St Louis, MO 63198
USA

J E Fry
Agriculture Group of Monsanto
700 Chesterfield Parkway North
St Louis MO 63198
USA

I Godwin
Department of Agriculture
University of Queensland Qld 4072
Australia

P Gras
CSIRO
Division of Plant Industry
Grain Quality Research Laboratory
North Ryde NSW 2113
Australia

F C Greene
USDA
Russell Regional Research Centre
PO Box 5677
Athens GA 30613
USA

H P Guan
Department of Biochemistry
Michigan State University
East Lansing
MI 48824
USA

F Gubler
Co-operative Research Centre for Plant Science
G P O Box 475
Canberra ACT 2601
Australia

R B Gupta
CSIRO
Division of Plant Industry
GPO Box 1600
Canberra ACT 2601
Australia

I A Haak
Queensland Wheat Research Institute
P O Box 2282
Toowoomba Qld 4350
Australia

N G Halford
Department of Agricultural Sciences
AFRC Institute of Arable Crops Research
University of Bristol
Long Ashton Research Station
Long Ashton
Bristol BS18 9AF
UK

N Harris
Department of Botany
University of Durham
Science Laboratories
South Road
Durham DH1 3LE
UK

D G He
Department of Biochemistry
University of Queensland 4072
Australia

F Heidekamp
Center for Phytotechnology
RUL/TNO
Department of Molecular Plant Biotechnology
Wassenaarseweg 64
2333 AL Leiden
The Netherlands

R J Henry
Queensland Agricultural Biotechnology Centre
Gehrmann Laboratories
University of Queensland Qld 4072
Australia

S Hoekstra
Center for Phytotechnology
RUL/TNO
Department of Molecular Plant Biotechnology
Wassenaarseweg 64
2333 AL Leiden
The Netherlands

M Iwata
Meiji Seika Ltd.
Pharmaceutical Research Centre
Japan

J Jacobsen
CSIRO
Division of Plant Industry
GPO Box 1600
Canberra ACT 2601
Australia

K K Kartha
Plant Biotechnology Institute
National Research Council
110 Gymnasium Place
Saskatoon
Saskatchewan
Canada S7N 0W9

S Karunaratne
Department of Biochemistry
University of Queensland 4072
Australia

G M Kishore
Plant Sciences Technology
The Agricultural Group
The Monsanto Company
St Louis MO 63198
USA

L. Ko
Queensland Agricultural Biotechnology Centre
Gehrmann Laboratories
University of Queensland 4072
Australia

M Kreis
Universite de Paris-Sud
Biologie du Developpement des Plantes
Batiment 430
F-91400
Orsay Cedex
France

P. Lafiandra
Department of Agrobiology and Agrochemistyr
University of Tuscia
Viterbo
Italy

Y Libal-Weksler
Department of Biochemistry
Michigan State University
East Lansing
MI 48824
USA

D McElroy
CSIRO
Division of Plant Industry
GPO Box 1600
Canberra ACT 2601
Australia

G E McKinnon
Queensland Agricultural Biotechnology Centre
Gehrmann Laboratories
University of Queensland Qld 4072
Australia

N. Margiotta
Germplasm Institute
C.N.R.
Bari
Italy

F van der Mark
Center for Phytotechnology
RUL/TNO
Department of Molecular Plant Biotechnology
Wassenaarseweg 64
2333 AL Leiden
The Netherlands

C Morris
National Food Authority
P O Box 7186
Canberra MC ACT 2610
Australia

A Mouradov
Department of Biochemistry
University of Queensland 4072
Australia

E Mouradova
Department of Biochemistry
University of Queensland 4072
Australia

N S Nehra
Plant Biotechnology Institute
National Research Council
110 Gymnasium Place
Saskatoon
Saskatchewan
Canada S7N 0W9

C Nojiri
Meiji Seika Ltd.
Pharmaceutical Research Centre
Japan

T W Okita
Institute of Biological Chemistry
Washington State University
Pullman
WA 99164
USA

S Ooba
Faculty of Agriculture
Gifu University
Japan

M E Pajeau
Agriculture Group of Monsanto
700 Chesterfield Parkway North
ST Louis MO 63198
USA

W P Pawlowski
Department of Agronomy and Plant Genetics
University of Minnesota
St Paul MN 55108
USA

J Preiss
Department of Biochemistry
Michigan State University
East Lansing
MI 48824
USA

J Peacock
CSIRO
Division of Plant Industry
GPO Box 1600
Canberra ACT 2601
Australia

P H Quail
University of California
Berkeley/U.S.D.A.
Plant Genome Expression Centre
USA

H W Rines
US Department of Agriculture
Plant Science Research Unit
St Paul MN 55108
USA

P K Samarajeewa
Institute of Molecular and Cellular Biosciences
University of Tokyo
Japan

P R Shewry
Department of Agricultural Sciences
AFRC Institute of Arable Crops Research
University of Bristol
Long Ashton Research Station
Long Ashton
Bristol BS18 9AF
UK

M N Sivak
Department of Biochemistry
Michigan State University
East Lansing
MI 48824
USA

D A Somers
Department of Agronomy and Plant Genetics
University of Minnesota
St Paul MN 55108
USA

C Smith
National Food Authority
P O Box 7186
Canberra MC ACT 2610
Australia

V Srivastava
Laboratory of Plant Cell and Molecular Biology
1143 Fifield Hall
University of Florida
Gainsville, FL 32611-0692
USA

D Stark
Plant Sciences Technology
The Agricultural Group
The Monsanto Company
St Louis MO 63198
USA

A Tam
USDA Western Regional Research Centre
800 Buchanan Street
Albany CA
USA

A S Tatham
Department of Agricultural Sciences
AFRC Institute of Arable Crops Research
University of Bristol
Long Ashton Research Station
Long Ashton
Bristol BS18 9AF
UK

S Takamatsu
Fukui Agricultural Experiment Station
Japan

S Toki
Department of Biological Science
Faculty of Science
Hokkaido University
Sapporo 060
Japan

K A Torbet
Department of Agronomy and Plant Genetics
University of Minnesota
St Paul MN 55108
USA

H Uchimiya
Institute of Molecular and Cellular Biosciences
University of Tokyo
Bunkyo-ku, Tokyo 113
Japan

I K Vasil
Laboratory of Plant Cell and Molecular Biology
1143 Fifield Hall
University of Florida
Gainsville, FL 32611-0692
USA

V Vasil
Laboratory of Plant Cell and Molecular Biology
1143 Fifield Hall
University of Florida
Gainsville, FL 32611-0692
USA

R Williams
Department of Plant Biology
University of California
Berkeley CA 94720
USA

B Witrzens
CSIRO
Division of Plant Industry
GPO Box 1600
Canberra ACT 2601
Australia

Y Wan
Department of Plant Biology
University of California
Berkeley CA 94710
USA

J T Weeks
USDA Western Regional Research Centre
800 Buchanan Street
Albany CA
USA

C W Wrigley
CSIRO
Division of Plant Industry
Grain Quality Research Laboratory
North Ryde NSW 2113
Australia

R Wu
Department of Biochemistry
Molecular and Cell Biology,
Cornell University
Ithaca NY 14853
USA

D Xu
Department of Biochemistry
Molecular and Cell Biology,
Cornell University
Ithaca NY 14853
USA

Y M Yang
Department of Biochemistry
University of Queensland 4072
Australia

W Zhang
Department of Biochemistry
Molecular and Cell Biology,
Cornell University
Ithaca NY 14853
USA

M van Zijderveld
Royal Sluis
Westeinde 161-163
BM Enkhuizen
The Netherlands

INDEX